DMV Seminar
Band 10

Springer Basel AG

Jürgen Jost

Nonlinear Methods in Riemannian and Kählerian Geometry

Delivered at the German Mathematical
Society Seminar
in Düsseldorf in June, 1986

Revised 2nd edition

1991

Springer Basel AG

Author

Jürgen Jost
Mathematisches Institut
Ruhr-Universität Bochum
Universitätsstrasse 150
D–4630 Bochum 1

The seminar was made possible through the support of the
Stiftung Volkswagenwerk

Deutsche Bibliothek Cataloging-in-Publication Data

Jost, Jürgen:
Nonlinear methods in Riemannian and Kählerian geometry:
delivered at the German Mathematical Society seminar in
Düsseldorf in June, 1986 / Jürgen Jost. – 2., rev. ed.

(DMV-Seminar ; Bd. 10)
ISBN 978-3-0348-7708-4   ISBN 978-3-0348-7706-0 (eBook)
DOI 10.1007/978-3-0348-7706-0
NE: Deutsche Mathematiker-Vereinigung: DMV-Seminar

© 1991 Springer Basel AG
Originally published by Birkhäuser Verlag Basel in 1991

ISBN 978-3-0348-7708-4

Preface

In this book, I present an expanded version of the contents of my lectures at a Seminar of the DMV (Deutsche Mathematiker Vereinigung) in Düsseldorf, June, 1986. The title "Nonlinear methods in complex geometry" already indicates a combination of techniques from nonlinear partial differential equations and geometric concepts. In older geometric investigations, usually the local aspects attracted more attention than the global ones as differential geometry in its foundations provides approximations of local phenomena through infinitesimal or differential constructions. Here, all equations are linear. If one wants to consider global aspects, however, usually the presence of curvature leads to a nonlinearity in the equations. The simplest case is the one of geodesics which are described by a system of second order nonlinear ODE; their linearizations are the Jacobi fields. More recently, nonlinear PDE played a more and more prominent rôle in geometry. Let us list some of the most important ones:

- harmonic maps between Riemannian and Kählerian manifolds

- minimal surfaces in Riemannian manifolds

- Monge-Ampère equations on Kähler manifolds

- Yang-Mills equations in vector bundles over manifolds.

While the solution of these equations usually is nontrivial, it can lead to very significant results in geometry, as solutions provide maps, submanifolds, metrics, or connections which are distinguished by geometric properties in a given context. All these equations are elliptic, but often parabolic equations are used as an auxiliary tool to solve the elliptic ones.

In this book, we first develop the geometric framework and then derive the equations that will subsequently be discussed. We try to present a rather broad geometric context, sometimes with sidetracks and occasionally deriving a formula in local coordinates through classical tensor notation as well as in invariant notation. Of course, the size of the present book makes it impossible to provide proofs of all relevant geometric results. Our main purpose rather is to lead a reader who either knows the basic geometric results or is willing to take them for granted or look them up in the references provided by us, to interesting topics and methods of the current research in geometry. We also give a short survey of some necessary results about linear elliptic and parabolic PDE. We aim at a readership of analysts who want to learn about the relevance and analytic problems of PDE in geometry as well as of geometers who want to become familiar with nonlinear analysis as a tool in geometry. We should point out, however, that our presentation is selective and by no means complete. Nevertheless, as already mentioned, we develop a general framework which should also be useful as a reference and discuss some significant and typical techniques.

Let us now describe the remaining contents in some detail. First of all, here, we do not treat minimal surfaces, and the Monge-Ampère equations for Kähler-Einstein metrics are only derived, but not solved (This is done in the companion volume by Siu appearing in the same series). Harmonic maps and Yang-Mills equations are quite thoroughly discussed, and we present a solution of the harmonic map problem for a nonpositively curved image manifold by a parabolic method (the result of Al'ber and Eells-Sampson). As a motivation, we first give a proof of the Hodge theorem by a parabolic method, due to Milgram-Rosenbloom. We also give a solution of the parabolic Hermitian Yang-Mills equation which arises in the investigation of stable vector bundles over Kähler manifolds (see again Siu's volume in this regard). This is originally due to Donaldson, but our proof is much simpler than his. In the final chapter, we discuss geometric applications of harmonic maps. We first treat the fundamental group of nonpositively curved manifolds and then present Siu's results on the rigidity of Kähler manifolds, saying that under suitable curvature conditions, the topological type of a compact Kähler manifold already determines its complex structure. These results were generalized to the noncompact finite volume case by Yau and the author (cf. the items [JY3], [JY4] in the bibliography), but this requires in addition concepts and methods of a quite different nature, and therefore is omitted here.

I have made no attempt to compile a complete bibliography; a reader interested in contributions not mentioned here should check the bibliographies of the more recent relevant papers listed in our bibliography.

I would like to thank Gerd Fischer for organizing the Seminar and inviting me as a speaker, the participants for their interest in my lectures, and Yum-Tong Siu for a very stimulating cooperation in delivering these lectures.

I am particularly grateful to Tilmann Wurzbacher who very carefully checked my manuscript and pointed out several corrections and improvements. The high quality of the typoscript is the work of Armin Köllner to whom I also want to extend my thanks.

I also acknowledge financial support from SFB 256 (University of Bonn).

Preface to the 2nd edition

This edition coincides with the first one, except for updating the bibliography and for the correction of some inaccuracies and misprints. I am grateful to Jürgen Eichhorn and Xiao-Wei Peng for relevant comments.

Table of contents

1. Geometric Preliminaries

In this chapter, we assemble some basic material concerning connections on Riemannian, complex, and Kähler manifolds, and derive the nonlinear partial differential equations dealt with in this book, namely the harmonic map and Yang-Mills equations.

We also provide the necessary geometric background. In this chapter, we freely use standard material as presented in [G], [KN], [W], [GH], [FU], [EL], [J2], [Kl], usually without explicit reference to these sources.

1.1. Connections

Let M be a differentiable manifold. A bundle over M is given by a total space E, a fiber F, and a projection $\pi : E \to M$ so that each $x \in M$ has a neighbourhood U for which $E_{|U} = \pi^{-1}(U)$ is diffeomorphic to $U \times F$ with preserved base point, i.e. there is a diffeomorphism $\varphi : \pi^{-1}(U) \to U \times F$ with $\pi = p_1 \circ \varphi$, where $p_1 : U \times F \to U$ is the projection onto the first factor. In other words, the bundle is locally trivial. Globally, however, a bundle is in general nontrivial, i.e. not diffeomorphic to the product $M \times F$.

Let $\{U_\alpha\}$ be an open covering of M, and let $\{\varphi_\alpha\}$ be the corresponding local trivializations. If $U_\alpha \cap U_\beta \neq \emptyset$, we get transition functions

$$\varphi_{\beta\alpha} : U_\alpha \cap U_\beta \to \mathrm{Diff}(F)$$

via

$$\varphi_\beta \circ \varphi_\alpha^{-1}(x, v) = (x, \varphi_{\beta\alpha}(x)v) \qquad (v \in F)$$

We have

$$\varphi_{\alpha\alpha}(x) = \mathrm{id}_F \qquad\qquad (x \in U_\alpha)$$

$$\varphi_{\alpha\beta}(x)\varphi_{\beta\alpha}(x) = \mathrm{id}_F \qquad\qquad (x \in U_\alpha \cap U_\beta)$$

$$\varphi_{\alpha\gamma}(x)\varphi_{\gamma\beta}(x)\varphi_{\beta\alpha}(x) = \mathrm{id}_F \qquad\qquad (x \in U_\alpha \cap U_\beta \cap U_\gamma)$$

E can be reconstructed from its local transition functions, i.e. $E = \coprod_\alpha U_\alpha \times F_{/\sim}$ with

$$(x, v) \sim (y, w) :\Longleftrightarrow x = y \quad \text{and} \quad w = \varphi_{\beta\alpha}(x)v \qquad (x \in U_\alpha, \quad y \in U_\beta, \quad v, w \in F)$$

A section of E is a (smooth) map $f : M \to E$ with $\pi \circ f = \mathrm{id}$. The space of sections of E is denoted by $C^\infty(E)$ or $\Gamma(E)$.

We are interested only in two cases, namely where F is either a vector space V or a Lie group G, and we require that the transition functions respect the corresponding structures. This means that the transition functions are not allowed to have arbitrary values in $\mathrm{Diff}(F)$ but are required to assume values in a specified finite dimensional Lie group G which is called the structure group of the bundle.

A vector bundle, then, has as a fiber a (real or complex) vector space V of real dimension n and as structure group $GL(n, \mathbb{R})$ or a subgroup of it. For a principal bundle, we denote the total space by P, and the fiber is a Lie group G; the structure group is G itself or a subgroup of it, acting on G as a group of left translations. Right multiplication on G induces via the local trivializations an action of G on P from the right,

$$P \times G \to P, \qquad (p \cdot g)h = p \cdot gh$$

This action is free ($p \cdot g = p \Longleftrightarrow g = e$), and $\pi : P \to M$ is given by identifying each $x \in M$ with an orbit of this action, $\pi : P \to P_{/G} = M$.

The groups G of interest for us are $GL(n, \mathbb{R})$, $O(n)$ and $SO(n)$, $U(n)$, $SU(n)$. As groups acting on a vector space, they are the transformations preserving a linear, Euclidean, or Hermitian structure. In general, an additional structure that has to be preserved leads to a restriction on the admissible transformations, or, as one says, to a reduction of the structure group. This can be seen both from the point of view of principal and of vector bundles, and we shall see that these two notions are only two different ways of looking at the same situation.

Namely, given a principal G-bundle $P \to M$ and a vector space V on which G acts (from the left), we construct the associated vector bundle E with fiber V as follows:

We have a right action of G on $P \times V$ which is free:

$$P \times V \times G \to P \times V, \qquad (p, v) \cdot g = (p \cdot g, g^{-1}v)$$

The projection $P \times V \to P \to M$ is invariant under this action, and

$$E = P \times_G V = P \times V_{/G} \to M$$

is a vector bundle with fiber $G \times_G V = G \times V_{/G} = V$ and structure group G. The transition functions for P also give transition functions for E via the left action of G on V.

Conversely, given a vector bundle E with structure group G, we construct a principal G-bundle P as

$$\coprod_\alpha U_\alpha \times G_{/\sim}$$

with

$$(x_\alpha, g_\alpha) \sim (x_\beta, g_\beta) :\Longleftrightarrow x_\alpha = x_\beta \in U_\alpha \cap U_\beta \quad \text{and} \quad g_\beta = \varphi_{\beta\alpha}(x) g_\alpha$$

($\{U_\alpha\}$ a local trivialization of E with transition functions $\varphi_{\beta\alpha}$).

P is considered as the bundle of admissible bases of E. In a local trivialization, each fiber of E is identified with $\mathbb{R}^n$ or $\mathbb{C}^n$, and each admissible basis is represented by a matrix contained in G. The transition functions then describe a change of basis. Also, the action of G on P is given by matrix multiplication from the right in each local trivialization.

The local trivializations allow to identify nearby fibers, but not in a canonical way. We now want to specify a way to relate fibers intrinsically. This is called a connection, covariant derivative, or parallel transport. (All these notions are equivalent.)

We start by explaining the notion of a connection in a principal G-bundle $\pi : P \to M$. Later on, we shall present several equivalent notions of a connection in a vector bundle and a reader who prefers to work with vector rather than with principal bundles can safely skip the discussion of connections in principal bundles and rather look at the notion of a covariant derivative below.

For each $p \in P$, in the tangent space $T_p P$ we have the subspace V_p, the tangent space at the fiber $\pi^{-1}(\pi(p))$, i.e. $V_p = \{X \in T_p P : \pi_* X = 0\}$. V_p is called the vertical space at p. A subspace H_p of $T_p P$ with

(1.1.1) $$V_p \oplus H_p = T_p P$$

is called a horizontal space. Although at each p, the vertical space V_p is canonically defined, there is no canonical choice of a horizontal space. A connection is a selection of a horizontal space at each $p \in P$ in a G-invariant way. i.e. we require (1.1.1) and

$$H_{pg} = (R_g)_* H_p \qquad \text{for all } g \in G,$$

where R_g is the right action of $g \in G$ on P $(R_g(p) = p \cdot g)$.

Eqivalently, we can choose for each p a projection $Q_p : T_p P \to V_p$ (in a G-invariant way) and put $H_p := \ker(Q_p)$. On the other hand, we recall that via the right action of G on

P, each fiber of P is identified with G, and hence each tangent space V_p to a fiber can be identified with the Lie algebra $\mathfrak{g}$ of G. This identification again is canonical, namely the identification between $\mathfrak{g} = T_e G$ and V_p is given by the differential of the embedding

$$i_p : G \to P, \qquad g \mapsto p \cdot g.$$

In particular, we obtain a trivialization of the vertical bundle. Hence, the family of projections Q_p can be identified with an element $\omega \in C^\infty (T^* P \otimes \mathfrak{g})$ with the properties

(1.1.2) $\qquad \omega((i_p)_* T) = T \qquad$ for $T \in \mathfrak{g} \qquad$ (i.e. ω is the identity on vertical vectors)

and

(1.1.3) $$(R_g)^* \omega = \mathrm{Ad}(g^{-1})\omega$$

i.e.

$$\omega((R_g)_* X) = \mathrm{Ad}(g^{-1})\omega(X) \qquad\qquad \text{for } X \in C^\infty (TP)$$
$$= (R_g)_* \omega(X)$$

(i.e. ω is equivariant).

(Here, Ad is the adjoint representation of G in $\mathfrak{g}$. Every automorphism φ of G induces an automorphism of $\mathfrak{g}$. $\mathfrak{g}$ can be considered as the space of left invariant vector fields on G, and if $A \in \mathfrak{g}$, then $\varphi_* A$ is also left invariant, and $\varphi_*[A, B] = [\varphi_* A, \varphi_* B]$. In particular, $I(g) : G \to G$, $h \mapsto ghg^{-1}$ induces an automorphism of $\mathfrak{g}$, denoted by Ad g. We have $(\mathrm{Ad}\ g)A = (R_{g^{-1}})_* A$, since $ghg^{-1} = R_{g^{-1}} L_g h$, and $(L_g)_* A = A$ as A is left invariant).

Let a connection be given in P.

We now want to define the horizontal lift of a curve and parallel transport along a curve. First of all, if $X \in T_x M$ and $p \in P$ with $\pi(p) = x$, then there is a unique $\tilde{X} \in H_p$ with $\pi_*(\tilde{X}) = X$; $\tilde{X}$ is the horizontal lift of X to p. if now $\gamma : [a, b] \to M$ is a smooth curve ($[a, b] \subset \mathbb{R}$), we get an induced connection on the pulled back bundle $\gamma^* P \to [a, b]$. We look at the horizontal lifts of the vector field $\frac{\partial}{\partial t}$, obtaining a horizontal vector field in

$T(\gamma^* P)$. Its maximal integral curves can be defined on $[a, b]$, and each image of such a maximal integral curve is called a horizontal lift of γ. For each $p \in \pi^{-1}(\gamma(a))$, there is a unique horizontal lift of γ through p, and following horizontal lifts, for each $t \in [a, b]$, we get a diffeomorphism between the fibers $\pi^{-1}(a)$ and $\pi^{-1}(t)$. This diffeomorphism is called parallel transport along γ from $\gamma(a)$ to $\gamma(t)$.

Let now E be a vector bundle with structure group G; the associated principal G-bundle P is considered as the bundle of admissible bases; if $G = \mathrm{GL}(n, \mathbb{R})$, $O(n)$, $U(n)$, we have linear, orthogonal, or unitary bases, resp. A horizontal lift of $\gamma : [a, b] \to M$ then is obtained by parallel transport of a basis along γ, and a vector X in the fiber of E over $\gamma(a)$ is then parallel transported along γ by requiring that its transport has constant coefficients w.r.t. the transport of the basis. In particular, if we have an orthogonal structure, i.e. $G=O(n)$, then this structure is preserved by parallel transport in the sense that the length of a vector and the angle between two vectors remain constant under parallel transport. Since parallel transport gives a way of identifying different fibers along a curve, this also allows to differentiate a section of E in the direction of the tangent vector of this curve. This process is called covariant differentiation. Parallel sections are considered to be constant, i.e. have vanishing derivative. Formally, the covariant derivative is an operator

$$D : C^\infty(E) \to C^\infty(E \otimes T^* M)$$

If $\sigma \in C^\infty(E)$, $X \in T_x M$, $\gamma : [0, 1] \to M$ a curve with $\gamma(0) = x$, $\dot{\gamma}(0) = X$, $\{e_i(t)\}$ the parallel transport of an admissible base $\{e_i(0)\}$ of E in $\gamma(0)$ along γ, $\sigma(\gamma(t)) = a_i(t)e_i(t)$ $(a_i \in C^\infty([0, 1], \mathbb{R}))$, then

$$D\sigma(X) := D_X \sigma := a_i'(0)e_i(0)$$

satisfies

$$D_{X+Y}\,\sigma = D_X\,\sigma + D_Y\,\sigma$$

(1.1.4)
$$D_{f\cdot X}\,\sigma = f D_X\,\sigma \qquad\qquad (f \in C^\infty(M,\mathbb{R}))$$

$$D_X\,(\sigma + \tau) = D_X\,\sigma + D_X\,\tau$$

$$D_X\,(f\sigma) = X(f)\cdot\sigma + f D_X\,\sigma$$

In particular, D is tensorial in X, but not in σ. It depends only on the value of X at x, but requires knowledge of σ in an (infinitesimal) neighbourhood of x.

If $G=O(n)$, then we have on each fiber of E a Euclidean scalar product $< \cdot,\cdot >$, and this remains invariant under parallel transport. If $\{e_i(t)\}$ is a parallel orthonormal base along γ, i.e. $< e_i(t), e_j(t) > = \delta_{ij}$, and if we have sections $\sigma = a_i(t)e_i(t)$, $\tau = b_i(t)e_i(t)$ along γ, $X = \dot\gamma(0)$ as before, then

$$X < a_i e_i, b_j e_j > = X(a_i b_i) = X(a_i)\cdot b_i + X(b_i)\cdot a_i$$
$$= < X(a_i)e_i, b_j e_j > + < a_i e_i, X(b_j)e_j >$$
$$= < D_X\,(a_i e_i), b_j e_j > + < a_i e_i, D_X\,(b_j e_j) >,$$

i.e. we have the Leibniz rule

(1.1.5)
$$d < \sigma,\tau > = < D\sigma,\tau > + < \sigma,D\tau >$$

where d denotes the exterior derivative.

Similarly, for $G=U(n)$, we have a corresponding Leibniz rule, and if $G=SU(n)$ and $\sigma_1,\ldots,\sigma_n$ are local orthonormal sections along γ, then

(1.1.6)
$$D_X\,(\sigma_1 \wedge \ldots \wedge \sigma_n) = 0$$

(here, of course, D is supposed to come from an $SU(n)$-connection)

If a trivialization of $TM \to M$ is given on U_α by coordinate vector fields $\frac{\partial}{\partial x^1},\ldots,\frac{\partial}{\partial x^m}$, and a trivialization of $E \to M$ by linearly independent sections $\mu_1,\ldots,\mu_n$, we put

$$D_{\frac{\partial}{\partial x^i}}\mu_j = \Gamma^k_{ij}\mu_k$$

If $X(t) = \gamma'(t)$, and $\mu(t) = a_k(t)\mu_k(\gamma(t))$ is a section of E along γ, then

$$(1.1.7) \qquad D_{X(t)}\mu(t) = a'_k(t)\,\mu_k(\gamma(t)) + \Gamma^k_{ij}(\gamma(t))\,\gamma'_i(t)\,a_j(t)\,\mu_k(\gamma(t))$$

(where γ has coordinate functions $(\gamma_1,\ldots,\gamma_m)$ w.r.t. the coordinate system $(x_1,\ldots,x_m)$).

We see that

$$(1.1.8) \qquad D_{X(t)}\mu(t) = 0$$

is a first order differential equation along γ which can be uniquely solved for a given initial value $\mu(0) \in \pi^{-1}(\gamma(0))$. If $\mu(t)$ solves (1.1.8), then $\mu(t)$ is the parallel transport of $\mu(0)$ along γ, or in other words $\mu(t)$ describes the horizontal lift of γ through $\mu(0)$. We conclude that a covariant derivative D satisfying (1.1.4) conversely determines the connection. Thus, connection, parallel transport, and covariant derivative are all equivalent notations.

The first term in (1.1.7) is completely independent of the choice of connection and determined by the exterior derivative. Hence, we write locally on U_α

$$D = d + A_\alpha$$

where $A_\alpha \in C^\infty(\mathfrak{g} \otimes T^*M_{|U_\alpha})$ is defined on U_α. $\mathfrak{g}$ here is considered as a subspace of $\mathrm{End}(V_x)$, the space of endomorphisms of the fiber, and it follows from the general construction that for a G-connection, A_α takes values in $\mathfrak{g}$ provided the local chart respects the G-structure. This can also be seen directly. By the Gram-Schmidt process, we can find bundle charts $\varphi_\alpha : E|_{U_\alpha} \to U_\alpha \times \mathbb{C}^n$ such that for each $x \in U_\alpha$, $e_j(x) := \varphi_\alpha^{-1}(x, e_j)$ constitutes a unitary basis of E_x where $e_1, \ldots, e_n$ is a unitary basis of $\mathbb{C}^n$. Then $de_j(x) = 0$ as e_j is constant in our chart. If now $x \in U_\alpha$, $X \in T_x M$,

$$0 = X\langle e_i, e_j \rangle = \langle A_\alpha(X)e_i, e_j \rangle + \langle e_i, A_\alpha(X)e_j \rangle$$

and $A_\alpha(X)$ is a skew hermitian matrix, i.e. an element of the Lie algebra $\mathfrak{u}(n)$ of $\mathrm{U}(n)$.

We can already deduce from the local formula (1.1.7) that A_α is given by matrix multiplication. The transformation rule can be determined as follows:

If $\varphi_{\beta\alpha} : U_\alpha \cap U_\beta \to G$ is the transition function, i.e. $\sigma_\beta = \varphi_{\beta\alpha}\sigma_\alpha$ for local sections σ_α, σ_β on U_α, U_β resp., then

$$\varphi_{\beta\alpha}(d + A_\alpha)\sigma_\alpha = (d + A_\beta)\sigma_\beta$$

and consequently

$$A_\alpha = \varphi_{\beta\alpha}^{-1} \, d\varphi_{\beta\alpha} + \varphi_{\beta\alpha}^{-1} \, A_\beta \, \varphi_{\beta\alpha}$$

In particular, the difference of two connections transforms as a tensor, and the space of all connections is an affine space. The difference is a 1-form with values in Ad E, i.e. an element of $\Omega^1(\text{Ad } E)$. Ad E is the bundle with fiber $(\text{Ad } E)_x \subset \text{End}(V_x)$ the space of those endomorphisms of V_x that lie in $\mathfrak{g}$; also Ad $E = P \times_G \mathfrak{g}$, where G acts on $\mathfrak{g}$ via the adjoint action.

Let $\Omega^i = C^\infty(\Lambda^i T^* M)$ be the space of i-forms on M. The exterior derivative defines the so-called de Rham complex

$$\Omega^0 \xrightarrow{d} \Omega^1 \xrightarrow{d} \ldots \xrightarrow{d} \Omega^n$$

Given a covariant derivative D, we construct

(1.1.9) $$\Omega^0(E) \xrightarrow{D} \Omega^1(E) \xrightarrow{D} \ldots \xrightarrow{D} \Omega^n(E)$$

where $\Omega^i(E) = C^\infty(E \otimes \Lambda^i T^* M)$, i.e. the space of E-valued i-forms, and

$$D(\sigma \otimes \omega) := D\sigma \wedge \omega + \sigma \otimes d\omega$$

for $\omega \in \Omega^i$, $\sigma \in C^\infty(E)$.

The curvature of D is defined as

$$F := D^2 : \Omega^0(E) \to \Omega^2(E)$$

It measures the failure of (1.1.9) to be a complex. We have locally

$$\begin{aligned}
F_\alpha(\sigma) &= (d + A_\alpha) \circ (d + A_\alpha)\sigma \\
&= (d + A_\alpha)(d\sigma + A_\alpha \sigma) \\
&= (dA_\alpha)\sigma - A_\alpha \, d\sigma + A_\alpha \, d\sigma + A_\alpha \wedge A_\alpha \sigma
\end{aligned}$$

i.e.

(1.1.10) $$F_\alpha = dA_\alpha + A_\alpha \wedge A_\alpha$$

In local coordinates $(x^1, \ldots, x^m)$, $A_\alpha = A_{\alpha,i}\, dx^i$ and

(1.1.11) $$F_\alpha = \left(\frac{1}{2} \left(\frac{\partial A_{\alpha,j}}{\partial x^i} - \frac{\partial A_{\alpha,i}}{\partial x^j} \right) + [A_{\alpha,i}, A_{\alpha,j}] \right)\, dx^i \wedge dx^j$$

Here, $[\cdot, \cdot]$ is the Lie algebra bracket, and each $A_{\alpha,i}$ is an element of $\mathfrak{g}$, considered as a subalgebra of $\text{End}(V_x)$. Thus, we can also write

$$F_\alpha = dA_\alpha + [A_\alpha, A_\alpha]$$

Of course, the product of A_α with itself has to be taken both for one-forms which leads to the notation $A_\alpha \wedge A_\alpha$, and in the Lie algebra $\mathfrak{g}$ as the Lie bracket, leading to the notation $[A_\alpha, A_\alpha]$. (Recall that locally A_α is a 1-form with values in $\mathfrak{g}$)

We compute

$$DF = dF + [A, F]$$

where the product again is formed in the Lie algebra as well as between the forms. On the other hand

$$dF = dA \wedge A - A \wedge dA$$

Consequently

(1.1.12) $$DF = dA \wedge A - A \wedge dA + A \wedge dA - dA \wedge A + [A, [A, A]]$$
$$= 0$$

This is the so-called (second) Bianchi identity.

(The vanishing of $[A, [A, A]]$ can best be seen in local coordinates:

In the preceding notation, this term becomes

$$(A_{\alpha,k} A_{\alpha,i} A_{\alpha,j} - A_{\alpha,k} A_{\alpha,j} A_{\alpha,i} - A_{\alpha,i} A_{\alpha,j} A_{\alpha,k} + A_{\alpha,j} A_{\alpha,i} A_{\alpha,k}) dx^k \wedge dx^i \wedge dx^j$$

Since $dx^k \wedge dx^i \wedge dx^j = dx^i \wedge dx^j \wedge dx^k$, the first and third term cancel each other, and likewise the second and fourth one.)

Furthermore, from (1.1.11) we see that F transforms via

$$(1.1.13) \qquad F_\alpha = \varphi_{\beta\alpha}^{-1} \, F_\beta \, \varphi_{\beta\alpha} = \mathrm{Ad}(\varphi_{\beta\alpha}^{-1}) \, F_\alpha$$

Hence, F transforms as a tensor and $F \in \Omega^2(\mathrm{Ad}\, E)$

(In order to understand the meaning of these operations, it may also be useful to stress the fact that A is (locally) a 1-form valued matrix; in the above notation, each $A_{\alpha,i}$ is a matrix. If we take a fixed local frame $e_1, \dots, e_n$ of E, we write $Ae_j = A_{jk}\, e_k$, where each A_{jk} now is a 1-form (and becomes a Christoffelsymbol Γ_{ij}^k when applied to $\frac{\partial}{\partial x^i}$).

If $e_1^*, \dots, e_n^*$ is a dual frame on E^*, i.e. $(e_i, e_j^*) = \delta_{ij}$, then we can determine the induced connection A^* on E^* via

$$0 = d(e_i, e_j^*) = (A_{ik}\, e_k, e_j^*) + (e_i, A_{jk}^*\, e_k)$$
$$= A_{ij} + A_{ji}^*$$

i.e.
$$A^* = -{}^tA$$

If we then have a local section $\sigma = \sigma_{ij}\, e_i \otimes e_j^*$ of $\mathrm{End}(E) = E \otimes E^*$, we compute

$$D(\sigma_{ij}\, e_i \otimes e_j^*) = d\sigma_{ij}\, e_i \otimes e_j^* + \sigma_{ij}\, A_{ik}\, e_k \otimes e_j^* - \sigma_{ij}\, A_{kj}\, e_i \otimes e_k^*$$
$$= d\sigma + [A, \sigma].$$

Thus, the induced connection on $\mathrm{End}(E)$, hence also on $\mathrm{Ad}(E)$, operates by taking the Lie bracket.)

In classical notation, the curvature tensor R is defined by F via

$$F : \Omega^0(E) \to \Omega^2(E)$$
$$\mu \mapsto R(\cdot, \cdot)\mu$$

We deduce from (1.1.11) that in local coordinates as above

$$(1.1.14) \qquad R\left(\frac{\partial}{\partial x^k}, \frac{\partial}{\partial x^l} \right) \mu_j = R_{jkl}^i \, \mu_i$$

with

$$(1.1.15) \qquad R_{jkl}^i = \frac{\partial \Gamma_{lj}^i}{\partial x^k} - \frac{\partial \Gamma_{kj}^i}{\partial x^l} + \Gamma_{lj}^m \, \Gamma_{km}^i - \Gamma_{kj}^m \, \Gamma_{lm}^i$$

and that

(1.1.16)
$$R(X,Y)\mu = D_X \, D_Y \, \mu - D_Y \, D_X \, \mu - D_{[X,Y]} \, \mu$$

Assuming for the sake of intuition $[X, Y] = 0$, we see that the curvature measures the local path dependence of parallel transport; the fact that parallel transport in general depends on the curve justifies our care in always speaking of parallel transport along a given curve. If the curvature vanishes, then parallel transport along two homotopic curves yields the same result; hence in this case parallel transport depends only on the homotopy class of the curve and thus yields a linear, orthogonal, or unitary (depending on G) representation of $\pi_1(M)$.

For the discussion of the Yang-Mills functional to follow, we also need the group of gauge transformations. A gauge transformation is a fiber preserving map $s : P \to P$ with $s(pg) = s(p)g$. s then is a cross section of $\mathrm{Aut}(E)$, the bundle with fiber G considered as automorphism group of the fiber V_x. In other words, $\mathrm{Aut}(E) = P \times_G G$, but here the action of G is by conjugation, so that $\mathrm{Aut}(E)$ is not a principal bundle. Since each fiber of $\mathrm{Aut}(E)$ is a group, we can multiply cross sections fiberwise.

We denote the group of gauge transformations by $\mathcal{G} = C^\infty(\mathrm{Aut}(E))$. $s \in \mathcal{G}$ acts on the space of connections via

$$s^*(D) = s^{-1} \circ D \circ s \qquad\qquad \text{for the covariant derivative } D$$

IF $\mu \in C^\infty(E)$, then consequently

$$s^*(D)\,\mu = s^{-1}\,D(s\mu)$$

In local formulae, putting $s(p) = s_\alpha(x) \cdot p$ for $p \in P_x$, the fiber of P over $x \in U_\alpha$, U_α coming from a trivialization of E as before, we get

(1.1.17)
$$s^*(A_\alpha) = s_\alpha^{-1}\,ds_\alpha + s_\alpha^{-1}\,A_\alpha\,s_\alpha,$$

where s_α now is a local section of $\mathrm{Aut}(E)$ and transforms via

$$s_\beta = \varphi_{\beta\alpha}\,s_\alpha\,\varphi_{\beta\alpha}^{-1},$$

using the same notation as when discussing the transformation behaviour of a connection. (Of course, this transformation rule is compatible with the group structure of $\mathcal{G}$.)

The action of s on the curvature F of a connection, namely

$$(1.1.18) \qquad\qquad s^*(F) = s^{-1} \circ F \circ s$$

($F \in C^\infty\left(\text{Ad } E \otimes \Lambda^2 T^* M\right)$, and the action takes place on the Ad E part, and is trivial on the $\Lambda^2 T^* M$ part), will be of particular importance.

Of special importance is the case where we have a connection ∇ on the tangent bundle TM of M. A curve $\gamma : [a, b] \to M$ is called a geodesic (w.r.t. ∇) if it is autoparallel in the sense

$$(1.1.19) \qquad\qquad \nabla_{\dot\gamma} \dot\gamma = 0,$$

i.e. its velocity field $\dot\gamma : [a, b] \to TM$ is parallel along γ.

In local coordinates $\left(\gamma(t) = \left(\gamma_1(t), \ldots, \gamma_m(t)\right)\right)$

$$\nabla_{\dot\gamma} \dot\gamma = \left(\ddot\gamma_k + \Gamma^k_{ij}(\gamma)\, \dot\gamma_i\, \dot\gamma_j\right) \frac{\partial}{\partial x^k} = 0$$

This is a system of ordinary second order equations, and consequently for each $x \in M$, $X \in T_x M$, there exists a unique maximal geodesic $\gamma_X : I_X \to M$ ($I_X \subset \mathbb{R}$, connected, $0 \in I_X$) with $\gamma_X(0) = x$, $\dot\gamma_X(0) = X$. $C := \{X \in TM : 1 \in I_X\}$ is an open starshaped neighbourhood of the zero section in TM. We define the exponential map

$$\exp : C \to M$$

via
$$\exp(X) = \gamma_X(1)$$

If $X \in T_x M$, $t \in [0, 1]$, then $\exp(tX) = \gamma_X(t)$. We conclude that the differential of $\exp : C \cap T_x M \to M$ at the origin $0 \in T_x M$ is the identity (identifying $T_0 T_x M$ with $T_x M$ in the canonical way).

1.2. Riemannian manifolds, geodesics, harmonic maps, and Yang-Mills fields

A Riemannian manifold is a smooth manifold M which is equipped with a Euclidean structure in each tangent space, varying smoothly from point to point. In other words, in each $T_x M$, there is an inner product $< \cdot, \cdot >$, and we can measure the length $|X| := < X, X >^{\frac{1}{2}}$ of a tangent vector, and the angle between $X, Y \in T_x M$. In local coordinates, we write

$$g_{ij} := < \frac{\partial}{\partial x^i}, \frac{\partial}{\partial x^j} >$$

We also write the metric as $g = g_{ij} \, dx^i \otimes dx^j$

As above, the requirement of compatibility with this Riemannian structure leads to a reduction of the structure group to $O(n)$, and corresponding connections ∇ necessarily satisfy $(X, Y, Z \in C^\infty(TM))$

$$(1.2.1) \qquad X < Y, Z > = < \nabla_X Y, Z > + < Y, \nabla_X Z >,$$

i.e. the metric is parallel in the sense that parallel transport preserves the inner product. A connection satisfying (1.2.1) is called Riemannian.

We define the torsion tensor of a connection ∇ on TM via $(X, Y \in C^\infty(TM))$

$$T(X, Y) := T_\nabla(X, Y) := \nabla_X Y - \nabla_Y X - [X, Y]$$

The requirement

$$(1.2.2) \qquad T \equiv 0$$

then expresses the compatibility between the connection and the Lie algebra structure on $C^\infty(TM)$.

On each Riemannian manifold (M, g), there is a unique connection ∇, the Levi-Civita connection, satisfying (1.2.1) and (1.2.2). It is given by

$$(1.2.3) \qquad < \nabla_X Y, Z > = \frac{1}{2} \Big\{ X < Y, Z > - Z < X, Y > + Y < Z, X > \\ - < X, [Y, Z] > + < Z, [X, Y] > + < Y, [Z, X] > \Big\}$$

In local coordinates, consequently the Christoffel symbols Γ_{ij}^k can be expressed through the metric (g_{ij}) and its derivatives, namely (putting $(g^{ij}) := (g_{ij})^{-1}$)

(1.2.4)
$$\Gamma_{ij}^k := \frac{1}{2}g^{kl}\left(\frac{\partial g_{jl}}{\partial x^i} + \frac{\partial g_{li}}{\partial x^j} - \frac{\partial g_{ij}}{\partial x^l}\right)$$

(This follows from (1.2.3), since $[\frac{\partial}{\partial x^i}, \frac{\partial}{\partial x^j}] = 0$ for all i,j.)

In the sequel, ∇ will always be the Levi-Civita connection on (M, g).

If $\gamma : [a, b] \to M$ is a (piecewise) smooth curve, then its length is defined as

$$L(\gamma) := \int_a^b < \dot{\gamma}(t), \dot{\gamma}(t) >^{\frac{1}{2}} dt = \int_a^b \|\dot{\gamma}(t)\| \, dt$$

and its energy as

$$E(\gamma) := \frac{1}{2}\int_a^b < \dot{\gamma}(t), \dot{\gamma}(t) > \, dt$$

The functional L determines the distance function $d(\cdot, \cdot) : M \times M \to \mathbf{R}^+$.

<u>Lemma:</u> *The geodesics w.r.t. the Levi-Civita connection are precisely the stationary points (w.r.t. fixed boundary values $\gamma(a), \gamma(b)$) of the energy functional. Geodesics satisfy $|\dot{\gamma}| \equiv$ const. Stationary points of the length functional also give rise to geodesics, but possibly only after reparametrization, as they need not to be parametrized proportionally to arclength, i.e. need not satisfy $|\dot{\gamma}| \equiv$ const.*

<u>pf.:</u> Let $\gamma : [a, b] \to M$ and consider a smooth variation

$$\gamma(\cdot, \cdot) : [a, b] \times (-\varepsilon, \varepsilon) \to M$$

with $\gamma(\cdot, 0) = \gamma(\cdot)$, $\gamma(a, s) = \gamma(a)$, $\gamma(b, s) = \gamma(b)$ for all $s \in (-\varepsilon, \varepsilon)$
Let $\dot{\gamma}(t, s) := X(t, s) := d\gamma(t, s)\frac{\partial}{\partial t}$, $Y(t, s) := d\gamma(t, s)\frac{\partial}{\partial s}$, $a \leq t \leq b$,
$-\varepsilon < s < \varepsilon$.
Then

$$\frac{d}{ds}L(\gamma(\cdot,s)) = \int_a^b \frac{d}{ds} <X,X>^{\frac{1}{2}} dt \qquad\qquad (X = X(t,s))$$

$$= \int_a^b \frac{1}{<X,X>^{\frac{1}{2}}} <\nabla_Y X, X> dt, \text{ since } \nabla \text{ is Riemannian}$$

$$= \int_a^b \frac{1}{<X,X>^{\frac{1}{2}}} <\nabla_X Y, X> dt \text{ , since } \nabla \text{ is torsion free}$$

$$= -\int_a^b <Y, \nabla_X \frac{X}{|X|}> dt \qquad\qquad \text{, since } \nabla \text{ is Riemannian}$$

$$\text{and } Y(a,s) = Y(b,s) = 0$$

If $\frac{d}{ds}L(\gamma(\cdot,s))|_{s=0} = 0$ for all variations, i.e. for all Y, then

$$\nabla_X \frac{X}{|X|} = 0 \qquad\qquad \text{at } s = 0,$$

i.e.

$$\nabla_{\dot\gamma} \frac{\dot\gamma}{|\dot\gamma|} = 0$$

We put

$$s(t) := \int_a^t |\dot\gamma(t)| dt$$

and denote by $t(s)$ the inverse function of $s(t)$. Then the preceding equation is equivalent to

$$\tilde\gamma(s) := \gamma(t(s))$$

being geodesic.

In the same way, critical points of the energy functional satisfy

$$\nabla_{\dot\gamma} \dot\gamma = 0 \qquad\qquad \text{, i.e. are geodesics}$$

Since

$$\frac{d}{dt} <\dot\gamma(t), \dot\gamma(t)> = 2 <\nabla_{\dot\gamma}\dot\gamma, \dot\gamma> = 0,$$

$|\dot\gamma| = $ const. for a geodesic, i.e. geodesics are parametrized proportionally to arclength.

$$\text{qed.}$$

Note that by elementary results about ODE, geodesics are automatically smooth. This can also be deduced by elementary considerations based on their length minimizing property. Although the concept of length seems geometrically more natural than the concept of energy, we prefer to work with the energy functional as here critical points automatically acquire a good parametrization.

If one knows the geodesics of a Riemannian manifold, one can also recover parallel transport as follows. First, for a vector field $X(t)$ along a geodesic $\gamma(t)$ to be parallel, its length $|X(t)|$ and its angle with $\dot{\gamma}(t)$, $< X(t), \dot{\gamma}(t) >$, have to be constant. Likewise, two parallel vector fields $X(t)$, $Y(t)$ along γ need to have constant product $< X(t), Y(t) >$. Finally, the condition that the (Levi-Civita) connection be torsion free prevents $X(t)$ from rotating around γ. For example, in $\mathbb{R}^3$, if γ is a geodesic in the x^3-direction, i.e. $\dot{\gamma} = \frac{\partial}{\partial x^3}$, then $X(t) := \cos\alpha(x^3)\frac{\partial}{\partial x^1} + \sin\alpha(x^3)\frac{\partial}{\partial x^2}$ cannot be parallel along γ, i.e. satisfy $\nabla_{\dot{\gamma}} X(t) = 0$, unless $\alpha(x^3) \equiv \text{const}$, since $[X(t), \dot{\gamma}(t)] = \frac{\partial\alpha}{\partial x^3}\left(\sin\alpha(x^3)\frac{\partial}{\partial x^1} - \cos\alpha(x^3)\frac{\partial}{\partial x^2}\right)$.

Furthermore, the parallel transport of $X \in T_{g(0)}M$ along an arbitrary curve $g(t)$ can be determined as follows. For each $\varepsilon > 0$, one finds points $0 = t_0 < t_1 < \ldots < t_N = 1$, $N = N(\varepsilon)$, with

$$d\big(g(t_{i-1}), g(t_i)\big) \leq \varepsilon \qquad (i = 1, \ldots, N).$$

One connects $g(t_{i-1})$ and $g(t_i)$ by a geodesic γ_i with $L(\gamma_i) \leq \varepsilon$ and transports X parallelly from $g(t_{i-1})$ to $g(t_i)$ along γ_i. $\varepsilon \to 0$ then yields parallel transport of X along g.

For $p \in M$, we let

$$\exp : T_p M \to M$$

be the exponential map for the Levi-Civita connection.

We define the injectivity radius of p as

$$i(p) := \sup\{r > 0 : \exp_{p|B(0,r)\subset T_p M} \text{ is injective}\}$$

and

$$i(M) := \inf_{p \in M} i(p).$$

If M is a compact Riemannian manifold, then $i(M) > 0$.

One also has

<u>Lemma 1.2.1:</u> *If one chooses an orthonormal basis* $e_1, \ldots, e_m$ *of* $T_p M$ *and identifies* $T_p M$
with $\mathbb{R}^m$ *via this basis, then*

$$\exp_p^{-1} : B(p, r) \to \mathbb{R}^m$$

$$(B(p, r) := \{q \in M : d(p, q) \leq r\}, \quad r < i(p))$$

defines local coordinates $(x^1, \ldots, x^m)$ *in* $B(p, r)$ *satisfying*

$$g_{ij}(p) = \; < \frac{\partial}{\partial x^i}, \frac{\partial}{\partial x^j} > (p) = \delta_{ij}$$

$$\left(\frac{\partial}{\partial x^k} g_{ij} \right)(p) = 0 \qquad\qquad (i, j, k = 1, \ldots, m)$$

These coordinates are called normal or exponential coordinates. Often, computations of coordinate invariant expressions are greatly facilitated by computing everything in normal coordinates. We shall encounter several examples of this.

Let us also assemble the basic facts about the curvature tensor of a Riemannian manifold, i.e. the curvature tensor of its Levi- Civita connection. R is given by

$$(1.2.5) \qquad\qquad R(X, Y)Z = \nabla_X \nabla_Y Z - \nabla_Y \nabla_X Z - \nabla_{[X,Y]} Z$$

for $X, Y, Z \in C^\infty(TM)$.

In local coordinates, cf. (1.1.14)

$$(1.2.6) \qquad\qquad R\left(\frac{\partial}{\partial x^k}, \frac{\partial}{\partial x^l} \right) \frac{\partial}{\partial x^j} = R^i_{jkl} \frac{\partial}{\partial x^i}$$

We put

$$R_{ijkl} := g_{im} R^m_{jkl},$$

i.e.

$$(1.2.7) \qquad\qquad R_{ijkl} = \; < R\left(\frac{\partial}{\partial x^k}, \frac{\partial}{\partial x^l} \right) \frac{\partial}{\partial x^j}, \frac{\partial}{\partial x^i} >$$

R_{ijkl} satisfies

(1.2.8) $$R_{ijkl} = -R_{jikl} = -R_{ijlk} = R_{klij}$$

(1.2.9) $$R_{ijkl} + R_{iklj} + R_{iljk} = 0 \qquad \text{(Bianchi's first identity)}$$

(1.2.10) $$\frac{\partial}{\partial x^h} R_{ijkl} + \frac{\partial}{\partial x^i} R_{jhkl} + \frac{\partial}{\partial x^j} R_{hikl} = 0 \qquad \text{(Bianchi's second identity,}$$

$$\text{cf. } (1.1.12))$$

If $A = a^i \frac{\partial}{\partial x^i}$, $B = b^i \frac{\partial}{\partial x^i} \in T_x M$ are linearly independent, then the sectional curvature of the plane spanned by A and B in $T_x M$ is defined to be

(1.2.11) $$K(A \wedge B) := \; < R(A, B)B, A > \cdot \frac{1}{|A \wedge B|^2}$$

$$= \frac{R_{ijkl} a^i b^j a^k b^l}{g_{ik} g_{jl} (a^i a^k b^j b^l - a^i a^j b^k b^l)}$$

The Ricci curvature in the direction $A \in T_x M$ is given by averaging the sectional curvatures of all planes containing A, i.e.

(1.2.12) $$\text{Ric}(A) = g^{jl} < R\left(A, \frac{\partial}{\partial x^j}\right) \frac{\partial}{\partial x^l}, A >$$

The Ricci tensor R_{ik} is then given by

(1.2.13) $$R_{ik} = g^{jl} R_{ijkl}$$

Likewise, the scalar curvature R is given by averaging all sectional curvatures at X, i.e.

(1.2.14) $$R = g^{ik} R_{ik}$$

A geodesic is a mapping from an interval into a Riemannian manifold which is critical for the energy functional. We now want to generalize this notion by replacing the interval by a general Riemannian manifold, defining the corresponding notion of energy, and determining the equations that critical points, the so-called harmonic maps have to satisfy.

Suppose that N and M are Riemannian manifolds of dimension n and m, resp., with metric tensors $(\gamma_{\alpha\beta})$ and (g_{ij}), resp., in some local coordinate charts $x = (x^1, \ldots, x^n)$ and

$f = (f^1, \ldots, f^m)$ on N and M, resp. Let $(\gamma^{\alpha\beta}) = (\gamma_{\alpha\beta})^{-1}$. If $f : N \to M$ is a C^1-map, we can define the energy density

$$(1.2.15) \qquad e(f) := \frac{1}{2}\gamma^{\alpha\beta}(x)\, g_{ij}(f)\, \frac{\partial f^i}{\partial x^\alpha} \frac{\partial f^j}{\partial x^\beta}$$

Intrinsically, the differential df of f, given in local coordinates by

$$df = \frac{\partial f^i}{\partial x^\alpha} dx^\alpha \otimes \frac{\partial}{\partial f^i}$$

can be considered as a section of the bundle $T^*N \otimes f^{-1}TM$. Then

$$(1.2.16) \qquad e(f) = \frac{1}{2}\gamma^{\alpha\beta} < \frac{\partial f}{\partial x^\alpha}, \frac{\partial f}{\partial x^\beta} >_{f^{-1}TM}$$

$$= \frac{1}{2} < df, df >_{T^*N \otimes f^{-1}TM}$$

i.e. $e(f)$ is the trace of the pullback via f of the metric tensor of M. In particular, $e(f)$ is independent of the choice of local coordinates and thus intrinsically defined.

(Note that the Riemannian structure of N is a scalar product on TN which then induces a scalar product on T^*N by duality. The product of $f^{-1}TM$ is just the pullback of the product on TM. Likewise, from $dx^\alpha \left(\frac{\partial}{\partial x^\beta}\right) = \delta_{\alpha\beta}$, hence $\frac{\partial}{\partial x^\gamma}\left(dx^\alpha\left(\frac{\partial}{\partial x^\beta}\right)\right) = 0$, we get the induced connection on T^*N :

$$\nabla_{\frac{\partial}{\partial x^\gamma}} dx^\alpha = -\Gamma^\alpha_{\beta\gamma}\, dx^\beta \qquad)$$

We define the energy of f as

$$E(f) := \int_N e(f)\, dN.$$

If f is of class C^2 and $E(f) < \infty$, and f is a critical point of E, then it is called *harmonic* and satisfies the corresponding Euler-Lagrange-equations. These are of the form

$$(1.2.17) \qquad \frac{1}{\sqrt{\gamma}} \frac{\partial}{\partial x^\alpha}\left(\sqrt{\gamma}\, \gamma^{\alpha\beta} \frac{\partial}{\partial x^\beta} f^i\right) + \gamma^{\alpha\beta}\, \Gamma^i_{jk} \frac{\partial}{\partial x^\alpha} f^j \frac{\partial}{\partial x^\beta} f^k = 0$$

in local coordinates, where $\gamma = \det(\gamma_{\alpha\beta})$ and the Γ^i_{jk} are the Christoffel symbols on M.

(1.2.17) is proved as follows. If f is critical, then for all admissible variations φ (e.g. $\varphi \in C_c^\infty(N)$, and $\varphi_{|\partial N} = 0$ if $\partial N \neq \emptyset$)

$$\frac{d}{dt} E(f + t\varphi)_{|t=0} = 0.$$

and thus

$$0 = \int_N \left(\gamma^{\alpha\beta}(x)\, g_{ij}(f(x)) \frac{\partial f^i}{\partial x^\alpha} \frac{\partial \varphi^j}{\partial x^\beta} + \frac{1}{2}\gamma^{\alpha\beta}\, g_{ij,k}\, \varphi^k \frac{\partial f^i}{\partial x^\alpha} \frac{\partial f^j}{\partial x^\beta} \right) \sqrt{\gamma}\, dx$$

$$= -\int_N \frac{\partial}{\partial x^\beta} \left(\sqrt{\gamma}\, \gamma^{\alpha\beta} \frac{\partial f^i}{\partial x^\alpha} \right) g_{ij}\, \varphi^j\, dx - \int_N \gamma^{\alpha\beta}(x) \frac{\partial f^i}{\partial x^\alpha} \frac{\partial f^k}{\partial x^\beta} g_{ij,k}\, \varphi^j \sqrt{\gamma}\, dx$$

$$+ \int_N \frac{1}{2}\gamma^{\alpha\beta}\, g_{ij,k}\, \varphi^k \frac{\partial f^i}{\partial x^\beta} \frac{\partial f^j}{\partial x^\alpha} \sqrt{\gamma}\, dx$$

since φ is compactly supported

and from this, putting $\eta^i = g_{ij}\varphi^j$, i.e. $\varphi^j = g^{jl}\eta^l$, and using the symmetry of $\gamma^{\alpha\beta}$ in the second integral,

$$0 = -\int_N \frac{\partial}{\partial x^\beta} \left(\sqrt{\gamma}\, \gamma^{\alpha\beta} \frac{\partial f^i}{\partial x^\alpha} \right) \eta^i\, dx - \int_N \frac{1}{2}\gamma^{\alpha\beta} g^{lj}(g_{ij,k} + g_{kj,i} - g_{ik,j}) \frac{\partial f^i}{\partial x^\alpha} \frac{\partial f^k}{\partial x^\beta} \eta^l \sqrt{\gamma}\, dx$$

which implies (1.2.17) by the fundamental lemma of the calculus of variations.

We can also consider this from an intrinsic point of view.

We let ψ be a vector field along f, i.e. a section of $f^{-1}(TM)$. In local coordinates

$$\psi = \psi^i(x) \frac{\partial}{\partial f^i},$$

and

$$d\psi = \nabla_{\frac{\partial}{\partial x^\alpha}} \left(\psi^i \frac{\partial}{\partial f^i} \right) dx^\alpha$$

$$= \frac{\partial \psi^i}{\partial x^\alpha} \frac{\partial}{\partial f^i} \otimes dx^\alpha + \psi^i\, \Gamma^k_{ij} \frac{\partial f^j}{\partial x^\alpha} \frac{\partial}{\partial f^k} \otimes dx^\alpha$$

then is a section of $T^*N \otimes f^{-1}(TM)$. ψ then induces a variation of f via

$$f_t(x) = \exp_{f(x)}(t\, \psi(x)).$$

We compute (noting $\nabla_{\frac{\partial}{\partial t}} df_t = \nabla_{\frac{\partial}{\partial x^\alpha}} \left(\frac{\partial f_t^i}{\partial t} \frac{\partial}{\partial f^i} \right) dx^\alpha = d\psi$, since $\frac{\partial}{\partial t}$ commutes with all derivatives w.r.t. x, and since the derivative of $\exp_p$ at $0 \in T_pM$ is the identity, cf. 1.1),

$$\frac{d}{dt} E(f_t)_{|t=0} = \int_N <df, d\psi>$$

$$= \int <df, \nabla_{\frac{\partial}{\partial x^\alpha}} \left(\psi^i \frac{\partial}{\partial f^i} \right) dx^\alpha >$$

$$= -\int <\nabla_{\frac{\partial}{\partial x^\alpha}} df, \psi^i \frac{\partial}{\partial f^i} dx^\alpha >$$

(in case N has a boundary, we have to require for this step - integration by parts - that ψ has compact support in the interior of N; in any case, of course ψ should be compactly supported)

$$= -\int < \text{trace } \nabla df, \psi >$$

Here, of course, ∇ is the covariant derivative in $T^*N \otimes f^{-1}TM$. In this notation, the Euler-Lagrange equations take the form

(1.2.18) $$\tau(f) := \text{trace } \nabla df = 0$$

We can also check directly that (1.2.17) and (1.2.18) are equivalent; namely

$$\nabla_{\frac{\partial}{\partial x^\beta}}(df) = \nabla_{\frac{\partial}{\partial x^\beta}}\left(\frac{\partial f^i}{\partial x^\alpha}\right) dx^\alpha \frac{\partial}{\partial f^i}$$

$$= \frac{\partial}{\partial x^\beta}\left(\frac{\partial f^i}{\partial x^\alpha}\right) dx^\alpha \frac{\partial}{\partial f^i} + \left(\nabla^{T^*N}_{\frac{\partial}{\partial x^\beta}} dx^\alpha\right) \frac{\partial f^i}{\partial x^\alpha} \frac{\partial}{\partial f^i} + \left(\nabla^{f^{-1}TM}_{\frac{\partial}{\partial x^\beta}} \frac{\partial}{\partial f^i}\right) \frac{\partial f^i}{\partial x^\alpha} dx^\alpha$$

$$= \frac{\partial^2 f^i}{\partial x^\alpha \partial x^\beta} dx^\alpha \frac{\partial}{\partial f^i} - {}^N\Gamma^\alpha_{\beta\gamma} dx^\gamma \frac{\partial f^i}{\partial x^\alpha} \frac{\partial}{\partial f^i} + {}^M\Gamma^k_{ij} \frac{\partial}{\partial f^k} \frac{\partial f^j}{\partial x^\beta} \frac{\partial f^i}{\partial x^\alpha} dx^\alpha$$

(here, we distinguish the Christoffel symbols of N and M by corresponding superscripts,) and thus since $\tau(f) = \text{trace } \nabla df$,

$$\tau^k(f) = \gamma^{\alpha\beta} \frac{\partial^2 f^k}{\partial x^\alpha \partial x^\beta} - \gamma^{\alpha\beta} {}^N\Gamma^\gamma_{\alpha\beta} \frac{\partial f^k}{\partial x^\gamma} + \gamma^{\alpha\beta} {}^M\Gamma^k_{ij} \frac{\partial f^i}{\partial x^\alpha} \frac{\partial f^j}{\partial x^\beta}.$$

We thus obtain a nonlinear elliptic system of partial differential equations, where the principal part is the Laplace–Beltrami operator on N and is therefore in divergence form, while the nonlinearity is quadratic in the gradient of the solution.

Remark: In the physical literature, the energy integral is called action.

Let V be a Euclidean vector space of dimension n, $e_1, \ldots, e_n$ an orthonormal basis of V, and denote by ΛV the exterior algebra of V. We define the star operator

$$* : \Lambda V \to \Lambda V$$

by

(1.2.19)
$$*(e_{i_1} \wedge \ldots \wedge e_{i_p}) = \pm e_{j_1} \wedge \ldots \wedge e_{j_{n-p}},$$

where $\{i_1, \ldots, i_p, j_1, \ldots, j_{n-p}\}$ is a permutation of $\{1, \ldots, n\}$, and we use the $+$ (-) sign, if the permutation is even (odd). One checks that

$$** : \Lambda^p V \to \Lambda^p V$$

satisfies

(1.2.20)
$$** = (-1)^{np+p} \text{ id}$$

Now let M be a compact, oriented, m-dimensional Riemannian manifold, and let as before $\Omega^i = C^\infty(\Lambda^i T^* M)$. Then $(x \in M)$

$$* : \Lambda^p T_x^* M \to \Lambda^{m-p} T_x^* M$$

is determined by the Riemannian metric (inducing a Euclidean structure on $T_x^* M$) and the orientation of M (determining the sign in (1.2.19)), and we obtain

$$* : \Omega^p \to \Omega^{m-p}$$

If as before, the metric is locally given by

$$g_{ij} = <\frac{\partial}{\partial x^i}, \frac{\partial}{\partial x^j}>, \qquad |x| = <x, x>^{\frac{1}{2}},$$

then

$$<dx^i, dx^j> = g^{ij} \qquad ((g^{ij}) = (g_{ij})^{-1})$$

for the induced inner product on $T^* M$.

From this we check

$$*(1) = (\det(g_{ij}))^{\frac{1}{2}} \, dx^1 \wedge \ldots \wedge dx^m = dM$$

is the volume form, and we define an inner product on Ω^p by $(\varphi, \psi \in \Omega^p)$

$$(\varphi, \psi) := \int_M \varphi \wedge *\psi$$

(this is welldefined since $\varphi \wedge *\psi$ is an m-form, and M is compact). It is easily checked that this product is symmetric and positive definite. (Actually, $\varphi \wedge *\psi = <\varphi, \psi> e_1 \wedge \ldots \wedge e_m$, where $<\varphi, \psi>$ is the induced product on p-forms.)

We can also complexify things, i.e. look at sections Φ, Ψ of $\Lambda^p T^* M \otimes \mathbb{C}$ and put

$$(\Phi, \Psi) := \int_M \Phi \wedge *\bar{\Psi}$$

which yields a Hermitian symmetric product.

If $d : \Omega^{p-1} \to \Omega^p$ is the exterior derivative, then we let d^* be its adjoint w.r.t. $(\cdot, \cdot)$, i.e. for $\varphi \in \Omega^{p-1}$, $\psi \in \Omega^p$

$$(d\varphi, \psi) = (\varphi, d^* \psi)$$

Using Stokes' Theorem and (1.2.20), one checks

(1.2.21) $$d^* = (-1)^{m+mp+1} * d* : \Omega^p \to \Omega^{p-1}$$

The Laplacian $\Delta^+ = dd^* + d^* d : \Omega^p \to \Omega^p$ then becomes a selfadjoint operator and satisfies

$$*\Delta^+ = (-1)^m \Delta^+ *$$

Note: All the preceding adjoints are adjoints only in a formal sense, as we do not worry about the maximal spaces on which our operators are welldefined.

If the metric of M is given in local coordinates by $g_{ij} dx^i \otimes dx^j$, we put $(g^{ij}) := (g_{ij})^{-1}$ and $g := \det(g_{ij})$.

The Laplace-Beltrami operator

$$\Delta^- = -\Delta^+ : \Omega^0 \to \Omega^0$$

is then given by

$$\Delta^- = \frac{1}{\sqrt{g}} \frac{\partial}{\partial x^i} \left(\sqrt{g} \, g^{ij} \frac{\partial}{\partial x^j} \right)$$

in these coordinates. (This operator was already encountered before as the principal part in (1.2.17).)

(As a side remark, let us also relate Δ^- to the divergence and gradient operators on a Riemannian manifold.

For a function f on M, we put

$$\nabla f := \operatorname{grad} f := g^{ij} \frac{\partial f}{\partial x^i} \frac{\partial}{\partial x^j} \qquad \text{in local coordinates,}$$

i.e. for a vector field X

$$< \operatorname{grad} f, X > = Xf = df(X),$$

so that grad f is the adjoint of df w.r.t. $< \cdot, \cdot >$. Next, for vector fields X, Z,

$$\operatorname{div} Z := \operatorname{trace} (X \mapsto \nabla_X Z).$$

In local coordinates

$$\operatorname{div} Z = \frac{1}{\sqrt{g}} \frac{\partial}{\partial x^j} \left(\sqrt{g}\, Z^j \right) \qquad \text{with } g := \det(g_{ij}).$$

This follows most easily from Gauss' Theorem:

Let φ be a function with compact support inside a coordinate chart U of M; then, with $dV = \sqrt{g}\, dx^1 \wedge \ldots \wedge dx^m$ being the volume element of M,

$$\int_U \varphi \left(\operatorname{div} Z - \frac{1}{\sqrt{g}} \frac{\partial}{\partial x^j} \left(\sqrt{g}\, Z^j \right) \right) dV$$
$$= \int_U \left(\varphi \cdot \operatorname{div} Z + \frac{\partial \varphi}{\partial x^j} Z^j \right) dV = \int_U \operatorname{div} (\varphi Z)\, dV = 0,$$

and since this holds for any such φ, the above formula for div Z follows.

Then

$$\Delta^- f = \frac{1}{\sqrt{g}} \frac{\partial}{\partial x^i} \left(\sqrt{g}\, g^{ij} \frac{\partial f}{\partial x^j} \right) = \operatorname{div} \operatorname{grad} f.)$$

Let now E be a vector bundle over M with structure group $SU(n)$, M as before a compact, oriented Riemannian manifold.

For $A, B \in \mathfrak{su}(n)$, we put

(1.2.22) $$A \cdot B := -\operatorname{tr}(A\, B)$$

which is nothing but the negative of the Killing form of $\mathfrak{su}(n)$ and yields a (positive definite) scalar product. Using the scalar product on $\Lambda^2 T_x^* M$ induced from the Riemannian metric, we obtain an inner product on the fibers of the bundle $\mathrm{Ad}\ E \otimes \Lambda^2 T^* M$, again denoted by $< \cdot, \cdot >$, and $|\cdot| = < \cdot, \cdot >^{\frac{1}{2}}$. For a connection D on E, we define the Yang-Mills functional as

$$YM(D) := \int_M < F_D, F_D > dM = \int_M |F_D|^2\ dM$$

where F_D is the curvature of D.

In order to compute the corresponding Euler-Lagrange equations, we look at variations of D of the form $D + tA$.

$$\begin{aligned} F_{D+tA}(\sigma) &= (D + tA)(D + tA)\sigma \\ &= D^2\sigma + tD(A\sigma) + tA \wedge D\sigma + t^2(A \wedge A)\sigma \\ &= \big(F + t(DA) + t^2(A \wedge A)\big)\ (\sigma)\ \ , \text{since } D(A\sigma) = (DA)\sigma - A \wedge D\sigma \end{aligned}$$

Thus

$$(1.2.23) \qquad \begin{aligned} \frac{d}{dt}YM(D + tA)_{|t=0} &= \frac{d}{dt}\int_M |F + tDA + t^2(A \wedge A)|^2\ dM_{|t=0} \\ &= 2\int < DA, F > dM \\ &= 2\int < A, D^*F > dM \end{aligned}$$

where again $D^* : \Omega^2(\mathrm{Ad}\ E) \to \Omega^1(\mathrm{Ad}\ E)$ is the adjoint of $D : \Omega^1(\mathrm{Ad}\ E) \to \Omega^2(\mathrm{Ad}\ E)$. The Euler-Lagrange equations for YM are thus

$$(1.2.24) \qquad\qquad\qquad\qquad D^* F_D = 0$$

which, by (1.2.21), is equivalent to

$$(1.2.25) \qquad\qquad\qquad\qquad D * F_D = 0$$

If F_D satisfies this equation, then D is called a Yang-Mills connection.
In local coordinates, if $g_{ij}(x) = \delta_{ij}$, then at x (1.2.25) becomes

$$(1.2.26) \qquad\qquad\qquad (D^*F)_j = \frac{\partial F_{ij}}{\partial x^i} + [A_i, F_{ij}] = 0$$

Since the product (1.2.22) is invariant under the adjoint action of G, and since $s \in \mathcal{G}$ acts on F via

$$s^*(F) = s^{-1}Fs$$

we see that YM and hence also (1.2.25) is invariant under the action of the infinite dimensional group $\mathcal{G}$ of gauge transformations. In particular, since the Yang-Mills equations have an infinite dimensional kernel, they are not elliptic. Therefore, one would like to divide out the action of $\mathcal{G}$. This is possible locally, but not globally.

In order to describe the precise result about local gauge fixing, we remember that the space of connections on E is an affine space, i.e. picking a base connection D_0,

$$\mathcal{A} = \{D = D_0 + A : A \in C^\infty (\mathrm{Ad}\, E \otimes T^*M)\}$$

Let $H^{k,p}$ be the Sobolev space of functions that have weak derivatives up to order k which are integrable to the pth power. Then

$$\mathcal{A}^{k,p} := \{D = D_0 + A, \ A \in H^{k,p}(M, \mathrm{Ad}\, E \otimes T^*M)\}$$

is the Sobolev space of $H^{k,p}$ connections. Because of the affine structure, this is independent of the choice of D_0.

The corresponding group of gauge transformations is

$$\mathcal{G}^{k+1,p} := H^{k+1,p}(M, \mathrm{Aut}(E))$$

(Note that $s \in \mathcal{G}^{k+1,p}$ acts on $\mathcal{A}^{k,p}$ via

$$A \mapsto s^{-1}D_0 s + s^{-1}As$$

so that we need one more derivative for s than for A. Using the Sobolev embedding theorem, one can see that for $k = 0$ or 1 and $(k+1)p > m(= \dim M)$, $\mathcal{G}^{k+1,p}$ is a smooth Lie group under pointwise multiplication, and the induced map

$$\mathcal{G}^{k+1,p} \times \mathcal{A}^{k,p} \to \mathcal{A}^{k,p}$$

is smooth).

Let $\|.\|_{k,p}$ denote a Sobolev $H^{k,p}$ norm. Then one has the local gauge fixing theorem of Uhlenbeck ([U2]):

<u>Thm.1.2.1:</u> *Let $M = B^m$ be the m-dimensional unit ball (since we are dealing with a local result, this is no restriction), $E = B^m \times \mathbb{R}^n$, G a compact subgroup of $SO(n)$, $2p \geq m$.*

$D = d + \tilde{A}$ with $\tilde{A} \in H^{1,p}(B^m, \mathbb{R}^m \times \mathcal{Y})$.

There exist $\kappa = \kappa(n) > 0$ and $c = c(n) < \infty$ with the property that if

$$\|F\|_{0,\frac{m}{2}}^{\frac{m}{2}} = \|d\tilde{A} + \tilde{A} \wedge \tilde{A}\|_{0,\frac{m}{2}}^{\frac{m}{2}} \leq \kappa$$

then D is gauge equivalent by an element $s \in H^{2,p}(B^m, G)$ to a connection $d + A$ (i.e. $A = s^{-1}ds + s^{-1}\tilde{A}s$) with

(i) $\qquad\qquad\qquad\qquad d^* A = 0$

(ii) $\qquad\qquad\qquad\qquad \|A\|_{1,p} \leq c\|F\|_{0,p}$

Note that for $2p = m$, the operations involved in a gauge transformation are not necessarily continuous.

This result implies regularity of connections $D \in \mathcal{A}^{1,p}$, $2p \geq m$, satisfying the Yang-Mills equations.

Namely, if

$$\int_M |F|^{\frac{m}{2}} \, dM = \|F\|_{0,\frac{m}{2}}^{\frac{m}{2}} < \infty$$

one can restrict to a small disk $B^m(0, \rho)$ with

$$\int_{B^m(0,\rho)} |F|^{\frac{m}{2}} \, dM \leq \kappa$$

Since this integral is invariant under dilations, one can control the size of ρ from below only if one knows $\|F\|_{0,p}^p$ with $p > \frac{m}{2}$ (using Hölder's inequality); knowledge of $\|F\|_{0,\frac{m}{2}}^{\frac{m}{2}}$ is not enough.

Thus applying the theorem, we get the Yang-Mills equation and (i), i.e. the system

$$d^* dA + A \wedge dA + A \wedge A \wedge A = 0$$

$$d^* A = 0$$

which can be combined into the elliptic system $(\Delta^+ = d^* d + dd^*)$

$$\Delta^+ A + A \wedge dA + A \wedge A \wedge A = 0$$

Then one can use elliptic regularity theory to derive the regularity of A, i.e. show that A is smooth (cf. 2.2 below). If $p > \frac{m}{2}$, one can obtain uniform estimates in terms of $\int |F|^p \, dM$.

One also has the global weak compactness theorem of Uhlenbeck ([U2]).

<u>Thm.1.2.2:</u> $p > \frac{m}{2}$. Let D_k be a sequence of $\mathcal{A}^{1,p}$ connections with $\int_M |F_{D_k}|^p \, dM \le c$ (independent of k). Assume M and G compact. Then, after selection of a subsequence, there exist gauge transformations $s_k \in \mathcal{G}^{2,p}$ for which $s_k^{-1} \circ D_k \circ s_k$ converges weakly in $\mathcal{A}^{1,p}$ to a connection D with

$$\int_M |F_D|^p \, dM \le c$$

If however G is Abelian, then the problem of gauge fixing becomes trivial. In this case, G acts trivially on Ad E, which itself is a trivial bundle, and $[\cdot, \cdot]$, the Lie algebra bracket in $\mathcal{U}$, vanishes identically.

Consequently, the curvature F remains invariant under the action of $\mathcal{G}$. We have then, since the bracket is trivial,

$$(1.2.27) \qquad\qquad F = dA$$

Likewise, the Bianchi identity becomes

$$(1.2.28) \qquad\qquad dF = 0,$$

and the Yang-Mills equation

$$(1.2.29) \qquad\qquad d^* F = 0.$$

Thus, for a Yang-Mills connection in the case of an Abelian G, the curvature F is a vector valued harmonic 2-form, since

$$(1.2.30) \qquad\qquad \Delta^+ F = 0 \Longleftrightarrow d^* F = 0 = dF,$$

and the existence and uniqueness of a solution follows from Hodge's Theorem, cf. 3.1. Still, the connection A is not unique, since only F, but not A remains invariant under the action of $\mathcal{G}$. If $s \in \mathcal{G}$ is of the form e^u, then

$$(1.2.31) \qquad\qquad s^*(A) = A + du$$

Writing (1.2.29) as

$$(1.2.32) \qquad\qquad d^* dA = 0,$$

one fixes the gauge by adding the equation

$$(1.2.33) \qquad\qquad d^* A = 0,$$

so that one has to solve

$$(1.2.34) \qquad\qquad \Delta^+ A = (d^* d + dd^*)A = 0.$$

In general, if A is a Yang-Mills connection, then so is

$$A + a \qquad, \ a \in \Omega^1(\mathrm{Ad}\ E)$$

with $da = 0$. If $H^1(M, \mathbb{R}) = 0$, we can choose $u \in \Omega^0(\mathrm{Ad}\ E)$ with $a = du$ (remember that $\mathrm{Ad}\ E$ is a trivial bundle); so that with $s = e^u$,

$$s^*(A) = A + a,$$

and consequently $\mathcal{G}$ acts transitively on the space of Yang-Mills connections.

Actually, in the present case, the only possible Abelian G is $U(1) = SO(2)$ (neglecting the trivial group SO(1)). In this case E is a unitary line bundle, Ad $E = M \times i\mathbf{R}$, and the curvature F of a connection represents $-2\pi i\, c_1(E)$ where $c_1(E)$ is the first Chern class of E, cf. 1.6 below.

Thus, for a Yang-Mills connection, the curvature F is the (unique) harmonic 2-form representing $-2\pi i\, c_1(E)$.

(For $G = U(1)$, one should write in (1.2.31) $s = e^{iu}$, with real valued u, hence $s^*(A) = a + i\, du$, so that the group of gauge transformations only introduces a phase factor)

In any case, Hodge theory, where one seeks a harmonic form representing a given cohomology class, represents a linear model for the nonlinear Yang-Mills problem.

We shall discuss an approach to Hodge theory in 3.1 that introduces some ideas that will also be useful for the nonlinear analysis.

In electromagnetic theory, one has a Lorentz manifold M instead of a Riemannian manifold but our formalism applies in the same way. The metric $(g_{\alpha\beta})_{\alpha,\beta=0,1,2,3}$ then has signature $(-,+,+,+)$. Local coordinates are denoted by $t = x^0, x^1, x^2, x^3$, where t represents the time direction. Given the electric field

$$E = (E_1, E_2, E_3)$$

and the magnetic field

$$B = (B_1, B_2, B_3),$$

one forms the electromagnetic field tensor

$$(F^\alpha_\beta)_{\alpha,\beta=0,1,2,3} = \begin{pmatrix} 0 & E_1 & E_2 & E_3 \\ E_1 & 0 & B_3 & -B_2 \\ E_2 & -B_3 & 0 & B_1 \\ E_3 & B_2 & -B_1 & 0 \end{pmatrix}$$

and

$$F_{\alpha\beta} = g_{\alpha\gamma} F^\gamma{}_\beta$$

The letter F here stands for Faraday, but we shall interpret F as a curvature tensor. Two of Maxwell's equations, namely the magnetostatic law

$$\operatorname{div} B = 0,$$

and the magnetodynamic law

$$\frac{\partial B}{\partial t} + \operatorname{curl} E = 0,$$

then take the form of Bianchi's identity

$$(1.2.35) \qquad\qquad dF = 0,$$

or in local coordinates

$$F_{\alpha\beta,\gamma} + F_{\beta\gamma,\alpha} + F_{\gamma\alpha,\beta} = 0.$$

These laws express the fact that there are no free magnetic poles and that the laws of electromagnetism are independent of the reference system.

The remaining equations of Maxwell, namely the electrostatic equation

$$\operatorname{div} E = 4\pi\rho \qquad\qquad (=\text{charge density})$$

and the electrodynamic one

$$\frac{\partial E}{\partial t} - \operatorname{curl} B = -4\pi I \qquad\qquad (=\text{current density})$$

can be written as

$$(1.2.36) \qquad\qquad d^* F = 4\pi * J$$

(with $J = (\rho, I_1, I_2, I_3)$), or again in local coordinates

$$F^{\alpha\beta}{}_{,\beta} = 4\pi J^\alpha \qquad\qquad (\alpha = 0,\ldots,3)$$

where $F^{\alpha\beta}$ is obtained from $F^\alpha{}_\beta$ in the standard way by index raising $(F^{\alpha\beta} = g^{\beta\gamma} F^\alpha{}_\gamma)$ If $J \equiv 0$, then we get the equation

$$(1.2.37) \qquad\qquad d^* F = 0,$$

and (1.2.35) and (1.2.37) are the same as (1.2.28), (1.2.29).

In terms of the vector potential $A = (A_0, A_1, A_2, A_3)$, B and E are given by

$$B = \text{rot}(A_1, A_2, A_3),$$
$$E = -\frac{\partial(A_1, A_2, A_3)}{\partial t} - \text{grad } A_0,$$

or

(1.2.38)
$$F = dA$$

and in local coordinates

$$F_{\alpha\beta} = A_{\beta,\alpha} - A_{\alpha,\beta}$$

From our point of view, A assumes the rôle of a connection, and a gauge transformation replaces A by

$$A + d\sigma$$

with an arbitrary function σ. Again, since $d^2 = 0$, this leaves F unchanged. Usually, one chooses a Lorentz gauge meaning that

$$d^* A = 0, \qquad\qquad \text{cf. (1.2.33)},$$

and the equations become

$$(d^* d + d d^*)A = \square^+ A = 4\pi J$$

where $\square^+$ is a generalization (up to sign) of the d'Alembert or wave operator, the hyperbolic analogue of the Laplace-Beltrami operator. In flat space-time (Minkowski space), this operator is

$$\square^- = -\frac{\partial^2}{\partial t^2} + \frac{\partial^2}{\partial x_1^2} + \frac{\partial^2}{\partial x_2^2} + \frac{\partial^2}{\partial x_3^2}.$$

Thus, although the formalism does not change when considering a Lorentz instead of a Riemannian manifold, the differential equation changes type, namely from elliptic to hyperbolic, and consequently quite a different sort of analysis is required for its solution.

One final comment: We already noted that Bianchi's identity (1.2.35) expresses the fact that the laws of physics are independent of the choice of reference system. Likewise, in general relativity, the Bianchi identity for the curvature tensor of the Lorentz metric expresses the fact that Einstein's field equations are independent of the choice of coordinates. Therefore, in general, the Bianchi identity $DF = 0$ expresses invariance under gauge transformations.

Historically, the preceding interpretation of Maxwell's equations is due to Hermann Weyl, in an attempt to combine Einstein's general relativity theory and electromagnetism into a single field theory.

His idea was that general relativity is naturally expressed in this framework through a structure group $O(4)$ whereas an electromagnetic field leads to an additional scale factor so that the combined structure group is $O(4) \times I\!R$. In the absense of charge, one therefore recovers the Levi-Civita connection, i.e. matter determines how a vector is rotated under parallel transport, whereas in the absence of matter, one gets an abelian theory, and charge determines how a vector is changed in length when transported around a curve, precisely as described above in the abelian case of gauge theory. This idea of change of scale, however, was rejected on physical grounds by Einstein, because if a clock, transported around a closed loop, would change its scale, it would also measure time differently, and hence the physics of a particle would depend on its history. Thus, every particle would have its own laws, and there would be no common physical laws, but universal chaos. Therefore, this idea was abandoned. If one considers $U(1)$ as gauge group, however (as it occurs in quantum mechanics), one would replace the gauge transformation e^u by e^{iu}, as noted above, so that instead of a scale factor, one would get a phase factor, and the above difficulties are gone. Anyway, Weyl's idea was later on revitalized in Yang-Mills theory, and gauge theories now play a crucial rôle in contemporary attempts to create a unified field theory for the four known basic forces of physics.

1.3. Jacobi fields and approximate fundamental solutions

Let $c(s,t) = c_t(s)$ be a family of geodesics parametrized by t. s usually will be taken as the arc length parameter on each geodesic. $J_t(s) = \frac{\partial}{\partial t}c(s,t)$ is then a Jacobi field. It satisfies the equation

$$(1.3.1) \qquad \nabla_{\frac{\partial}{\partial s}}\nabla_{\frac{\partial}{\partial s}} J_t - R\left(\frac{\partial c}{\partial s}, J_t\right)\frac{\partial c}{\partial s} = 0$$

which easily follows from $\nabla_{\frac{\partial}{\partial s}}\frac{\partial}{\partial s}c = 0$ and the definition of the curvature tensor.

From (1.3.1) we see that the tangential component of a Jacobi field J, $J^{tan} = <J, \frac{\partial c}{\partial s}> \frac{\partial c}{\partial s}$ satisfies

$$\nabla_{\frac{\partial}{\partial s}}\nabla_{\frac{\partial}{\partial s}} J^{tan} = 0$$

and is hence independent of the metric. In particular, J^{tan} is linear. In order to incorporate the tangential component in the estimates, we have to assume that we have curvature bounds

$$(1.3.2) \qquad \lambda \le K \le \mu, \qquad \lambda \le 0, \qquad \mu \ge 0$$

i.e. a nonpositive lower and a nonnegative upper bound, or else to assume $J^{tan} = 0$.

We need some definitions:

$'$ always denotes a derivative with respect to s, while $\dot{}$ is the differentiation with respect to t.

We put

$$c_\rho(s) = \begin{cases} \cos(\sqrt{\rho}s) & \text{if } \rho > 0 \\ 1 & \text{if } \rho = 0 \\ \cosh(\sqrt{-\rho}s) & \text{if } \rho < 0 \end{cases}$$

and

$$s_\rho(s) = \begin{cases} \frac{1}{\sqrt{\rho}}\sin(\sqrt{\rho}s) & \text{if } \rho > 0 \\ s & \text{if } \rho = 0 \\ \frac{1}{\sqrt{-\rho}}\sinh(\sqrt{-\rho}s) & \text{if } \rho < 0 \end{cases}$$

Both functions solve the Jacobi equation for constant sectional curvature ρ, namely

$$(1.3.3) \qquad f'' + \rho f = 0$$

with initial values $f(0) = 1$, $f'(0) = 0$, or $f(0) = 0$, $f'(0) = 1$, resp.

c will always be a geodesic arc parametrized by s proportionally to arclength, and usually $|c'| = 1$ for simplicity.

<u>Lemma 1.3.1:</u> Assume $K \leq \mu$ and $|c'| = 1$, and either $\mu \geq 0$ or $J^{\tan} \equiv 0$. Then we have

(1.3.4)
$$|J(0)|c_\mu(s) + |J|'(0)s_\mu(s) \leq |J(s)|$$

provided the left hand side is positive for all r with $0 < r < s$

<u>Lemma 1.3.2:</u> Assume $\lambda \leq K \leq \mu$, and either $\lambda \leq 0$ or $J^{\tan} \equiv 0$, $|c'| \equiv 1$, and in addition that $J(0)$ and $J'(0)$ are linearly dependent.

For a parameter τ, we define $f_\tau = |J(0)|c_\tau + |J|'(0)s_\tau$. If $s_{\frac{1}{2}(\lambda+\mu)} > 0$ on $(0, \rho)$, then

(1.3.5)
$$|J(s)| \leq |J(0)|c_\lambda(s) + |J|'(0)s_\lambda(s) ,$$

Proofs of the preceding estimates can be found in [BK] and [J2].

As an example of the application of Jacobi fields estimates to geometric constructions, we show

<u>Lemma 1.3.3:</u> Let $B(m, \rho) := \{x \in N : d(m, x) \leq \rho\}$ be a ball in some manifold N which is disjoint to the cut locus of its centre m. We assume for the sectional curvatures K in $B(m, \rho)$

$$-\omega^2 \leq K \leq \kappa^2 \qquad \text{and} \qquad \rho < \frac{\pi}{2\kappa}$$

We define $r(x) := d(x, m)$ and $f(x) := \frac{1}{2}d(x, m)^2$. Then $f \in C^2(B(m, \rho), \mathbb{R})$ and

(1.3.6)
$$|\operatorname{grad} f(x)| = r(x)$$

(1.3.7)
$$\kappa r(x) \operatorname{ctg}(\kappa r(x)) \cdot |v|^2 \leq D^2 f(v, v) \leq \omega r(x) \coth(\omega r(x)) \cdot |v|^2$$

for $x \in B(m, \rho)$ and $v \in T_x N$.

pf.: $\operatorname{grad} f(x) = -\exp_x^{-1} m$ which implies (1.3.6).

Since ∇df is diagonalizable, it suffices to show (1.3.7) for each eigenvector v of ∇df. Let $q(t)$ be a curve in N with $q(0) = x$ and $\dot{q}(0) = v$ and

$$c(s,t) = \exp_{q(t)}\left(s \, \exp_{q(t)}^{-1} m\right).$$

Then $\operatorname{grad} f(q(t)) = -\frac{\partial}{\partial s}\big|_{s=0} c(s,t)$, and hence

$$\nabla_v \operatorname{grad} f(x) = -\nabla_{\frac{\partial}{\partial t}} \frac{\partial}{\partial s} c(s,t)\big|_{s=0,t=0}$$
$$= -\nabla_{\frac{\partial}{\partial s}} \frac{\partial}{\partial t} c(s,t).$$

For fixed t, $J_t(s) = \frac{\partial}{\partial t} c(s,t)$ is the Jacobi field along the geodesic from m to $q(t)$ with $J_t(0) = \dot{q}(t)$ and $J_t(1) = 0 \in T_m N$. Hence $\nabla_v \operatorname{grad} f(x) = \nabla_{J_0(0)} \operatorname{grad} f(x) = -J_0'(0)$. Since

$$D^2 f(v,v) = \langle \nabla_v \operatorname{grad} f, v \rangle = -\langle J_0'(0), J_0(0) \rangle ,$$

(1.3.7) follows from (1.3.4) and (1.3.5) (since v is an eigenvector of ∇df, $J_0'(0) = -\nabla_v \operatorname{grad} f(x)$ and $J_0(0) = v$ are linearly dependent).

We now construct approximate fundamental solutions of the Laplace and the heat equation on manifolds.

Lemma 1.3.4: Let $B(m,\rho)$ be as in Lemma 1.3.3. $\Lambda^2 := \max(\kappa^2, \omega^2)$, and let Δ^- be the Laplace-Beltrami operator on N, and $n = \dim N$, $h(x) := d(x,m)^2$.

(1.3.8) $\qquad |\Delta^- \log r(x)| \le 2\Lambda^2 \qquad\qquad$ for $x \ne m$ if $n = 2$

(1.3.9) $\qquad |\Delta^- r(x)^{2-n}| \le \dfrac{n-2}{2}\Lambda^2 r^{2-n}(x) \qquad$ for $x \ne m$ if $n \ge 3$

and

(1.3.10) $\qquad \left|\left(\Delta^- - \dfrac{\partial}{\partial t}\right)\left(t^{\frac{-n}{2}} \exp\left(-\dfrac{h(x)}{4t}\right)\right)\right| \le 2\Lambda^2 \dfrac{h(x)}{4t} t^{\frac{-n}{2}} \exp\left(-\dfrac{h(x)}{4t}\right)$

$$\text{for } (x,t) \ne (m,0).$$

The proof follows through a straightforward computation from Lemma 1.3.3, since by Taylor expansion

$$(1 - \kappa^2 r^2)|v|^2 \leq D^2 f(v, v) \leq (1 + \omega^2 r^2)|v|^2 \qquad \text{for } v \in T_X N.$$

qed.

From the preceding lemma, one can derive approximate versions of Green's representation formula (for proofs, cf.[J2]).

Lemma 1.3.5: Let $B(m, \rho)$ be as above, $h(x) = d(x, m)^2$. Let ω_n denote the volume of the unit sphere in $\mathbb{R}^n$. If $\varphi \in C^2(B(m, \rho), \mathbb{R})$, then

(1.3.11) if $n = 2$

$$\left| \omega_2 \varphi(m) + \int_{B(m,\rho)} \Delta^- \varphi \cdot \log \frac{r(x)}{\rho} - \frac{1}{\rho} \int_{\partial B(m,\rho)} \varphi \right| \leq 2\Lambda^2 \int_{B(m,\rho)} |\varphi|$$

(1.3.12) if $n \geq 3$

$$\left| (n-2)\omega_n \varphi(m) + \int_{B(m,\rho)} \Delta^- \varphi \left(\frac{1}{r(x)^{n-2}} - \frac{1}{\rho^{n-2}} \right) - \frac{(n-2)}{\rho^{n-1}} \int_{\partial B(m,\rho)} \varphi \right|$$

$$\leq \frac{n-2}{2} \Lambda^2 \int_{B(m,\rho)} \frac{|\varphi|}{r(x)^{n-2}}.$$

We note that the error term is of lower order than the other two terms which are the same as in the Euclidean version of the Green representation formula.

In the parabolic case, one has

Lemma 1.3.6: Let $B(m, \rho)$ be as above and define

$$B(m, \rho, t_0, t) := \{(x, \tau) \in B(m, \rho) \times [t_0, t]\},$$

$$\varphi(\cdot, \tau) \in C^2(B(m, \rho), \mathbb{R}), \quad \varphi(x, \cdot) \in C^1([t_0, t], \mathbb{R}).$$

Then, with a constant c_n depending only on n

(1.3.13)

$$\left| (4\pi)^{\frac{n}{2}} \varphi(m,t) + \int_{B(m,\rho,t_0,t)} \left(\Delta^- - \frac{\partial}{\partial \tau} \right) \varphi(x,\tau)(t-\tau)^{-\frac{n}{2}} \right.$$

$$\left. \left\{ \exp \left(-\frac{r^2(x)}{4(t-\tau)} \right) - \exp \left(-\frac{\rho^2}{4(t-\tau)} \right) \right\} \, dx \, d\tau \right|$$

$$\leq \frac{c_n}{\rho^{n+2}} \int_{B(m,\rho,t_0,t)} |\varphi| + \frac{c_n}{\rho^{n+1}} \int_{\substack{r(x)=\rho \\ t_0 \leq \tau \leq t}} |\varphi(x,\tau)| + (t-t_0)^{-\frac{n}{2}} \int_{B(m,\rho)} |\varphi(x,t_0)| \, dx$$

$$+ 2\Lambda^2 \int_{B(m,\rho,t_0,t)} |\varphi(x,\tau)| \frac{r^2(x)}{t-\tau} (t-\tau)^{-\frac{n}{2}} \exp \left(-\frac{r^2(x)}{4(t-\tau)} \right) \, dx \, d\tau$$

and also

(1.3.14)

$$\left| (4\pi)^{\frac{n}{2}} \varphi(m,t) + \int_{B(m,\rho,t_0,t)} \left(\Delta^- - \frac{\partial}{\partial \tau} \right) \varphi(x,\tau)(t-\tau)^{-\frac{n}{2}} \right.$$

$$\left(\exp \left(-\frac{r^2(x)}{4(t-\tau)} \right) - \exp \left(-\frac{\rho^2}{4(t-\tau)} \right) \right) \, dx \, d\tau$$

$$\left. - \int_{B(m,\rho)} \varphi(x,t_0)(t-t_0)^{-\frac{n}{2}} \exp \left(-\frac{r^2(x)}{4(t-t_0)} \right) \, dx \right|$$

$$\leq \frac{c_n}{\rho^{n+2}} \int_{B(m,\rho,t_0,t)} |\varphi| + \frac{c_n}{\rho^{n+1}} \int_{\substack{r(x)=\rho \\ t_0 \leq \tau \leq t}} |\varphi(x,\tau)| + \frac{c_n}{\rho^n} \int_{B(m,\rho)} |\varphi(x,t_0)| dx$$

$$+ 2\Lambda^2 \int_{B(m,\rho,t_0,t)} |\varphi(x,\tau)| \frac{r^2(x)}{(t-\tau)} (t-\tau)^{-\frac{n}{2}} \exp \left(-\frac{r^2(x)}{4(t-\tau)} \right) \, dx \, d\tau.$$

1.4. Complex manifolds and vector bundles

A complex manifold M is a differentiable manifold with holomorphic coordinate transformations. A complex submanifold S of M is a submanifold which is a complex manifold itself. Locally, S can be described as the zero set of a collection $f_1, \ldots, f_k$ of holomorphic functions for which $\left(\frac{\partial f_i}{\partial z^j}\right)$ ($z^1, \ldots, z^m$ local holomorphic coordinates on M) has maximal rank, or as the image of an open set $U \subset \mathbb{C}^{m-k}$ under a holomorphic map $h : U \to M$ with $\operatorname{rank}(dh) = m - k$.

An analytic subvariety of M is a subset given locally as the zero set of a finite collection of holomorphic functions, without requiring any nondegeneracy condition. Consequently, an analytic subvariety V has smooth or regular points, around which V is locally a submanifold, and in general also singular points where this fails.

Let again $z^j = x^j + iy^j$, $j = 1, \ldots, m$, be local holomorphic coordinates near $z \in M$. $T_z^{\mathbb{R}} M = T_z M$ then is the ordinary (real) tangent space of M at z, and

$$T_z^{\mathbb{C}} M := T_z^{\mathbb{R}} M \otimes_{\mathbb{R}} \mathbb{C}$$

is the complexified tangent space

$$T_z^{\mathbb{C}} M = \mathbb{C} \left\{ \frac{\partial}{\partial z^i}, \frac{\partial}{\partial \bar{z}^i} \right\} = T_z' M \oplus T_z'' M$$

where $T_z' = \mathbb{C} \left\{ \frac{\partial}{\partial z^i} \right\}$ is the holomorphic and $T_z'' M = \mathbb{C} \left\{ \frac{\partial}{\partial \bar{z}^i} \right\}$ the antiholomorphic tangent space.

In $T_z^{\mathbb{C}} M$, we have the operation of conjugation, mapping $\frac{\partial}{\partial z^i}$ to $\frac{\partial}{\partial \bar{z}^i}$, $T_z'' M = \overline{T_z' M}$, and the projection $T_z^{\mathbb{R}} M \to T_z^{\mathbb{C}} M \to T_z' M$ is a $\mathbb{R}$-linear isomorphism.

A vector bundle E over a differentiable manifold M is called a complex vector bundle if each fiber is a complex vector space and the complex structure varies smoothly, i.e. we require local trivializations

$$\varphi : \pi^{-1}(U) \to U \times \mathbb{C}^k$$

for which the fiber $V_z = \pi^{-1}(z)$ is isomorphic to $z \times \mathbb{C}^k$.

For a complex vector bundle, in contrast to the definition of a complex manifold, one neither requires that the base M is complex nor that the transition functions

$$\varphi_{\beta\alpha} : U_\alpha \cap U_\beta \to GL(k,\mathbb{C})$$

are holomorphic.

If, however, M is a complex manifold and the transition functions are holomorphic, then we speak of a holomorphic vector bundle. An example of course is the holomorphic tangent bundle of a complex manifold.

On a complex manifold M, we can decompose Ω^k, the space of k-forms into subspaces $\Omega^{p,q}$ with $p + q = k$, where locally $\Omega^{p,q}$ is spanned by forms of the form

$$\varphi(z) \, dz^{i_1} \wedge \ldots \wedge dz^{i_p} \wedge d\bar{z}^{j_1} \wedge \ldots \wedge d\bar{z}^{j_p}$$

We can also define operators ∂ and $\bar{\partial}$ by defining its values on the preceding form as

$$\frac{\partial \varphi}{\partial z^i} \, dz^i \wedge dz^{i_1} \wedge \ldots \wedge d\bar{z}^{j_p}$$

and

$$\frac{\partial \varphi}{\partial \bar{z}^j} \, d\bar{z}^j \wedge dz^{i_1} \wedge \ldots \wedge d\bar{z}^{j_p} \qquad , \text{resp.}$$

Of course,

$$d = \partial + \bar{\partial}$$

One checks for example that

$$\bar{\partial}^2 = 0$$

Therefore, in the same way as d defines the de Rham cohomology, $\bar{\partial}$ gives rise to the Dolbeault cohomology groups $H_{\bar{\partial}}^{p,q}(M)$.

As we saw, there is no natural connection on a general vector bundle, but on a holomorphic vector bundle, we can define a natural $\bar{\partial}$ operator,

$$\bar{\partial} : \Omega^{p,q}(E) \to \Omega^{p,q+1}(E)$$

Namely, if $\{e^1, \ldots, e^n\}$ is a local holomophic frame, $\sigma \in \Omega^{p,q}(E)$, then locally

$$\sigma = \varphi_i \otimes e^i \qquad\qquad (\varphi_i \in \Omega^{p,q})$$

We put

$$\bar{\partial}\sigma = \bar{\partial}\varphi_i \otimes e^i$$

Since the transition functions are holomorphic, this is independent of the choice of frame. A Hermitian vector bundle is a complex vector bundle with a Hermitian inner product on each fiber depending smoothly on the base point. If the vector bundle is at the same time holomorphic, we have a Hermitian holomorphic vector bundle.

A Hermitian vector bundle has $U(n)$ as structure group. If the product on a fiber is denoted by $< \cdot, \cdot >$, a corresponding connection D has to be compatible with the Hermitian structure, i.e. has to satisfy

$$(1.4.1) \qquad d < \sigma, \tau > = < D\sigma, \tau > + < \sigma, D\tau >,$$

cf. (1.1.5).

If the bundle is holomorphic, then we can split $D = D' + D''$ with

$$D' : \Omega^0(E) \to \Omega^{1,0}(E)$$
$$D'' : \Omega^0(E) \to \Omega^{0,1}(E)$$

and the connection is compatible with the complex structure if

$$(1.4.2) \qquad D'' = \bar{\partial}$$

as defined above

Lemma 1.4.1: For a Hermitian holomorphic vector bundle E, there is a unique connection D compatible with both the Hermitian and the holomorphic structure, i.e. satisfying (1.4.1) and (1.4.2). We call D the metric complex connection for E.

Locally, if $\{e^i\}$ is a holomorphic frame,

$$h_{i\bar{j}} := < e^i, e^j >, \qquad (h^{i\bar{j}}) := (h_{i\bar{j}})^{-1}.$$

writing $D = d + A$, the matrix $A = (A_{ij})_{i,j=1,\ldots,n}$ is given by

$$(1.4.3) \qquad A_{ij} = (\partial h_{i\bar{k}})h^{j\bar{k}}$$

The curvature of D, $F = dA + A \wedge A$, reduces to $(h = (h_{i\bar{j}}))$

(1.4.4) $$F = \bar{\partial} A = \bar{\partial}\partial h \cdot h^{-1} + \partial h \cdot h^{-1} \wedge \bar{\partial} h \cdot h^{-1}$$

In particular, it is of type $(1,1)$.

Conversely, given a complex vector bundle E over a complex manifold M, the existence of a holomorphic structure on E is equivalent to the existence of a $\bar{\partial}$-operator

$$\bar{\partial} : \Omega^0(E) \to \Omega^{0,1}(E)$$

satisfying the product rule $(f \in C^\infty(M, \mathbb{C}), \sigma \in \Omega^0(E))$

$$\bar{\partial}(f\sigma) = \bar{\partial}f \cdot \sigma + f \, \bar{\partial}\sigma,$$

with

$$\bar{\partial}^2 = 0$$

(extended as $\bar{\partial} : \Omega^{0,1}(E) \to \Omega^{0,2}(E)$). If E carries a Hermitian metric, this in turn is equivalent to the existence of an $U(r)$-connection with curvature of type $(1,1)$. These assertions are a consequence of the Theorem of Newlander-Nirenberg, cf. [AHS].

Two $\bar{\partial}$-operators yield isomorphic holomorphic structures if they are conjugate by some $s \in C^\infty(GL(E))$ ($GL(E)$ is the bundle of complex linear automorphisms of E) ($\bar{\partial}_2 = s^{-1}\bar{\partial}_1 s$). Note that s need not be unitary.

If $h_1 = < \cdot, \cdot >_1$ and $h_2 = < \cdot, \cdot >_2$ are two Hermitian metrics with

$$< \sigma, \tau >_1 \; = \; < {}^t u \sigma, \tau >_2$$

for some positive definite, selfadjoint $u = h_1 h_2^{-1} \in C^\infty(GL(E))$, then the corresponding connections (cf. Lemma 1.4.1) are related by $(D_1 = \partial_1 + \bar{\partial}_1, D_2 = \partial_2 + \bar{\partial}_2)$

(1.4.5) $$\bar{\partial}_2 = \bar{\partial}_1$$

$$\partial_2 = u^{-1}\partial_1 u$$

and the curvatures $F_1, F_2 \in \Omega^{1,1}(\mathrm{Ad}\, E)$ by

(1.4.6) $$F_2 = F_1 + \bar{\partial}_1(u^{-1}\partial_1 u)$$

As shown in our discussion of connections, in the present case the group of gauge transformations is

$$\mathcal{G} = \{s \in C^\infty\left(\mathrm{GL}(E)\right) : \bar{s}^t s = \mathrm{id}\},$$

and a connection transforms via

(1.4.7)
$$D \to s^{-1} D s$$

with $D = \partial + \bar{\partial}$, we can write this as

(1.4.8)
$$\bar{\partial} \to s^{-1} \bar{\partial} s$$

$$\partial \to \bar{s}^t \partial \bar{s}^{t^{-1}} \qquad \text{(since } \bar{s}^t s = \mathrm{id})$$

Using this formula, we can extend this to an action of $\mathcal{G}^{\sigma} := C^\infty\left(\mathrm{GL}(E)\right)$. This action then no longer preserves the Yang-Mills functional, but two connections define isomorphic holomorphic structures, if and only if they are equivalent under this action of $\mathcal{G}^{\sigma}$.

If we want to identify the isomorphic holomorphic structures, we have to rewrite (1.4.8) as (D, is the transformed connection)

(1.4.9)
$$sD_{,}s^{-1} = \bar{\partial} + (s\bar{s}^t)\partial(s\bar{s}^t)^{-1}$$

and for the curvature

(1.4.10)
$$sF_{,}s^{-1} = F + \bar{\partial}\left((s\bar{s}^t)^{-1}\partial(s\bar{s}^t)\right)$$

so that with $u = s\bar{s}^t$ (selfadjoint and positive definite), we have the same situation as in (1.4.5), (1.4.6).

Chern classes of complex vector bundles

Let again E be a complex vector bundle over the compact manifold M, D a connection in E, $F = D^2 : \Omega^0(E) \to \Omega^2(E)$ its curvature; remember that in local trivializations, F transforms via

$$(1.4.11) \qquad F_\alpha = \varphi_{\beta\alpha}^{-1} F_\beta \varphi_{\beta\alpha} \qquad\qquad (\varphi_\beta = \varphi_{\beta\alpha}\varphi_\alpha) \qquad (\text{cf.}(1.1.13))$$

so that we can consider F as an element of $\Omega^2(\text{Ad } E)$. Since in the present case, the structure group is $\text{GL}(n,\mathbb{C})$, we have $\text{Ad } E = \text{End } E = \text{Hom}_{\mathbb{C}}(E; E)$.

A polynomial function P, defined on the space M_n of complex $n \times n$-matrices

$$P : M_n \to \mathbb{C}$$

which is homogeneous of degree k in its entries, is called invariant, if

$$P(A) = P(\varphi^{-1} A \varphi)$$

for all $A \in M_n$, $\varphi \in \text{GL}(n,\mathbb{C})$. For example, the elementary symmetric polynomials of the eigenvalues of A, denoted by $P^i(A)$, are invariant. They satisfy

$$\det(A + t \cdot \text{Id}) = \sum_{k=0}^{n} P^{n-k}(A) t^k$$

A k-linear form

$$\tilde{P} : M_n \times \ldots \times M_n \to \mathbb{C}$$

is called invariant, if for $A_1, \ldots, A_k \in M_n$, $\varphi \in \text{GL}(n,\mathbb{C})$

$$\tilde{P}(A_1, \ldots, A_k) = \tilde{P}(\varphi^{-1} A_1 \varphi, \ldots, \varphi^{-1} A_k \varphi)$$

An invariant k-form defines an invariant polynomial by restriction to the diagonal, i.e.

$$P(A) := \tilde{P}(A, \ldots, A)$$

Conversely, from an invariant polynomial, one can construct an invariant k-form by polarization $\left(\tilde{P}(A_1, \ldots, A_k) = \dfrac{(-1)^k}{k!} \sum_{j=1}^{k} \sum_{i_1 < \ldots < i_j} (-1)^j P(A_{i_1} + \ldots + A_{i_j}) \right).$

If now P is an invariant polynomial and F the curvature of D as above, then, because of (1.4.11),

$$P(F) := P(F_\alpha) \quad \text{locally}$$

is a welldefined global $2k$-form on M, i.e. independent of the local trivialization.

<u>Lemma 1.4.2</u>: Let P be an invariant polynomial of degree K. Then

    (i) $dP(F) = 0$

    (ii) The cohomology class $[P(F)] \in H^{2k}(M)$ is independent of the connection chosen for E.

    <u>pf.</u>: We let $\tilde{P}$ be an invariant k-form with $\tilde{P}(A, \ldots, A) = P(A)$. We extend D as an operator $D : \Omega^p(\text{End } E) \to \Omega^{p+1}(\text{End } E)$. Then for $B_i \in \Omega^{p_i}(\text{End } E)$

$$d\tilde{P}(B_1, \ldots, B_k) = \sum_i (-1)^{p_1 + \cdots + p_{i-1}} \tilde{P}(B_1, \ldots, DB_i, \ldots, B_k)$$

On the other hand, by the Bianchi identity

$$DF = 0,$$

and hence

$$dP(F) = 0$$

If D_0 and D_1 are two connections on E, we put $\eta = D_1 - D_0 \in \Omega^1(\text{End } E)$,

$$D_t := D_0 + t\eta$$

then is a family of connections with curvature

$$F_t = d(A + t\eta) + (A + t\eta) \wedge (A + t\eta)$$

where $D_0 = d + A$ as in 1.1.

Hence

$$\frac{\partial}{\partial t} F_t = D_t(\eta)$$

Thus

$$\frac{\partial}{\partial t} P(F_t) = k\tilde{P}(\frac{\partial}{\partial t} F_t, F_t, \ldots, F_t)$$
$$= k\tilde{P}(D_t \eta, F_t, \ldots, F_t)$$
$$= d(k\tilde{P}(\eta, F_t, \ldots, F_t)) \qquad \text{since } D_t F_t = 0 \text{ by}$$

Bianchi's identity

Consequently

$$P(F_1) - P(F_0) = \int_0^1 \frac{\partial}{\partial t} P(F_t) dt$$

is cohomologous to zero.

qed.

The lemma establishes the Weil homomorphism

$w :$ Algebra of graded invariant polynomials $\rightarrow H^{2*}(M)$

$w(P) = [P(F)]$

The Chern classes of E, then, are defined as

$$c_i(E) = \left[P^i\left(\frac{\sqrt{-1}}{2\pi} F\right) \right] \in H^{2i}(M),$$

where P^i is the ith elementary polynomial.

In other words,

$$(1.4.12) \qquad \det\left(\frac{\sqrt{-1}}{2\pi} F + t\mathrm{Id}\right) = \sum_{k=0}^{n} c_{n-k}(E) t^k$$

If M is a complex manifold, we define

$$c_i(M) := c_i(T'M)$$

Putting $x = t^{-1}$, we can rewrite (1.4.12) as

$$\sum_{i=0}^{n} c_i(E) x^i = \det\left(\frac{\sqrt{-1}}{2\pi} xF + \mathrm{Id}\right) =: \prod_{j=1}^{n} (1 + \lambda_j x),$$

where the λ_j are the eigenvalues of $\frac{\sqrt{-1}}{2\pi} F$ [1]. This is most convenient when one wants to express the Chern classes of bundles constructed from E; e.g.

$$\sum_{i=0}^{n} c_i(E^*)x^i = \prod_{j=1}^{n}(1 - \lambda_j x)$$

$$\sum_{i} c_i(\Lambda^p E)x^i = \prod_{1 \le j_1 < \ldots < j_p \le n}(1 + (\lambda_{j_1} + \ldots + \lambda_{j_1})x);$$

in particular

$$c_1(\Lambda^m T'M) = c_1(M) \qquad\qquad (m = \dim_{\mathbb{C}} M).$$

Also, for two bundle E, E' in obvious notation

$$\sum_{i=0}^{n+n'} c_i(E \oplus E')x^i = \prod_{j=1}^{n}(1 + \lambda_j x) \prod_{k=1}^{n'}(1 + \lambda'_k x)$$

$$\sum_{i=0}^{n \cdot n'} c_i(E \otimes E')x^i = \prod_{j,k}(1 + (\lambda_j + \lambda'_k)x)$$

---

[1] Note that each λ_j is a 2-form.

1.5 Kähler manifolds

We summarize the relevant results without proofs which can be found, e.g. in [W].

Suppose M is a differentiable manifold which has a complex as well as a Hermitian structure. Each of these structures defines a unique natural connection in the tangent bundle TM. The complex structure defines a connection compatible with the Hermitian metric and with the holomorphic structure in $T'M$. The Hermitian structure induces a Riemannian structure which defines the torsion free Levi-Civita connection in TM. Recall that there is a natural isomorphism between the real tangent bundle TM and the holomorphic tangent bundle $T'M$.

M is called a Kähler manifold if these two connections agree.

We write the Hermitian metric as

$$g_{i\bar{j}} \, dz_i \otimes d\bar{z}_j \qquad\qquad (g_{i\bar{j}} = g_{\bar{j}i} = \overline{g_{i\bar{j}}} = \overline{g_{\bar{j}i}})$$

(the corresponding Riemannian metric then is $2\mathrm{Re}\, g_{i\bar{j}} \, dz_i \otimes d\bar{z}_j$)

and define the Kähler form ω as

$$\omega = \frac{i}{2} \, g_{i\bar{j}} \, dz_i \wedge d\bar{z}_j$$

<u>Thm.1.5.1:</u> *Each of the following conditions is equivalent to M being Kähler*

(i) $d\omega = 0$

(ii) *in local coordinates* $\dfrac{\partial g_{i\bar{j}}}{\partial z^k} = \dfrac{\partial g_{k\bar{j}}}{\partial z^i}$ *for all* i, j, k

(or equivalently $\dfrac{\partial g_{i\bar{j}}}{\partial \bar{z}^l} = \dfrac{\partial g_{i\bar{l}}}{\partial \bar{z}^j}$*)*

(iii) *Locally, there exists a function F with* $g_{i\bar{j}} = \dfrac{\partial^2 F}{\partial z^i \partial \bar{z}^j}$

(F is called a Kähler potential), i.e. $\omega = \dfrac{i}{2} \partial \bar{\partial} F$

(iv) *At each point z_0, one can introduce holomorphic normal coordinates, i.e.*

$$g_{i\bar{j}}(z_0) = \delta_{ij}, \qquad \frac{\partial g_{i\bar{j}}}{\partial z^k}(z_0) = 0 = \frac{\partial g_{i\bar{j}}}{\partial \bar{z}^l}(z_0) \qquad\qquad \text{for all } i, j, k, l$$

Let us state some formulae in local coordinates on a Kähler manifold. First of all, for the inverse of $(g_{i\bar{j}})$, one uses the convention

$$g^{i\bar{j}}\, g_{k\bar{j}} = \delta_{ik} \qquad\qquad \text{(note the switch of indices)}$$

The Laplace-Beltrami operator becomes, with $g = \det(g_{i\bar{j}})$

$$\frac{1}{g}\frac{\partial}{\partial z^i}\left(g\, g^{i\bar{j}}\frac{\partial}{\partial z^{\bar{j}}}\right) = g^{i\bar{j}}\frac{\partial^2}{\partial z^i \partial z^{\bar{j}}}$$

The curvature tensor is given by

$$R_{i\bar{j}k\bar{l}} = \frac{\partial^2}{\partial z^{\bar{l}}\partial z^k}g_{i\bar{j}} - g^{s\bar{t}}\left(\frac{\partial}{\partial z^k}g_{i\bar{t}}\right)\left(\frac{\partial}{\partial z^{\bar{l}}}g_{s\bar{j}}\right).$$

We have

$$R_{ij\bar{k}\bar{l}} = R_{\bar{i}\bar{j}kl} = 0,$$

and therefore, with the first Bianchi identity

$$R_{i\bar{j}k\bar{l}} + R_{ik\bar{l}\bar{j}} + R_{i\bar{l}\bar{j}k} = 0$$

and

$$R_{i\bar{l}\bar{j}k} = -R_{i\bar{l}k\bar{j}},$$

we obtain

$$R_{i\bar{j}k\bar{l}} = R_{i\bar{l}k\bar{j}},$$

and likewise

$$R_{k\bar{j}i\bar{l}} = R_{i\bar{j}k\bar{l}}$$

With

$$\partial f := \frac{\partial f}{\partial z^i}dz^i, \qquad \bar{\partial} f = \frac{\partial f}{\partial z^{\bar{j}}}dz^{\bar{j}},$$

the Ricci form is obtained as

$$R_{k\bar{l}}\, dz^k \wedge dz^{\bar{l}} = -\partial\bar{\partial}\log\,\det(g_{i\bar{j}})$$

and the scalar curvature as

$$R = -\Delta^{\sim}\log\,\det(g_{i\bar{j}}).$$

(The minus sign arises as $\partial\bar\partial$ is a negative operator.)

For the Chern classes, we have for example

$$c_1(M) = \frac{i}{2\pi} \, g^{i\bar j} R_{i\bar j k\bar l} \, dz^k \wedge dz^{\bar l} = \frac{i}{2\pi} \, R_{k\bar l} \, dz^k \wedge dz^{\bar l}$$

where $(R_{k\bar l})$ is the Ricci tensor, and

$$c_2(M) = \frac{-1}{8\pi^2} \, g^{i\bar j} g^{k\bar l} \left(R_{i\bar j p q} R_{k\bar l s\bar t} - R_{i\bar l p q} R_{k\bar j s\bar t} \right) \, dz^p \wedge dz^q \wedge dz^s \wedge dz^{\bar t}$$

We recall the decomposition

$$\Omega^k = \sum_{p+k=k} \Omega^{p,q}$$

of k-forms into (p,q) forms, valid on a general complex manifold M. If ω is the Kähler form, we define

$$L : \Omega^{p,q} \to \Omega^{p+1,q+1}$$

via

$$L(\eta) = \eta \wedge \omega$$

We let

$$\Lambda = L^* : \Omega^{p,q} \to \Omega^{p-1,q-1}$$

where the adjoint is defined as above w.r.t. the product $(\varphi, \psi) = \int_M \varphi \wedge *\bar\psi$.

In particular

$$\Lambda(\omega) = 1$$

In general, for $\eta \in \Omega^{1,1}$

$$\Lambda\eta \cdot *(1) = \frac{1}{(m-1)!} \, \eta \wedge \omega^{m-1}$$

If

$$\omega = \frac{i}{2} \, g_{j\bar k} \, dz^j \wedge dz^{\bar k}, \qquad \eta = b_{i\bar j} \, dz^i \wedge dz^{\bar j}$$

then

(1.5.1) $$\Lambda\left(b_{i\bar j} \, dz^i \wedge dz^{\bar j} \right) = -2i \, g^{i\bar j} b_{i\bar j} = -2i \, \mathrm{tr}_g \eta$$

On a Kähler manifold, we have the following identities

(1.5.2)
$$[\Lambda, \bar{\partial}] = -i\, \partial^* \qquad\qquad ([A, B] := AB - BA)$$

(1.5.3)
$$[\Lambda, \partial] = i\, \bar{\partial}^*$$

(1.5.4)
$$\Delta^+ = dd^* + d^*d = 2(\bar{\partial}\bar{\partial}^* + \bar{\partial}^*\bar{\partial}) =: 2\Delta_{\bar{\partial}}$$
$$= 2(\partial\partial^* + \partial^*\partial) =: 2\Delta_{\partial}$$

In particular, Δ^+ maps $\Omega^{p,q}$ into itself.

(1.5.5)
$$[\Lambda, \Delta^+] = 0$$

We have the

Hodge Decomposition Theorem: For a compact Kähler manifold M

$$H^k(M, \mathbb{C}) = \bigoplus_{p+q=k} H^{p,q}(M)$$

$$H^{p,q}(M) = \overline{H^{q,p}(M)}$$

Hodge theory represents each element in $H^{p,q}(M)$ by a harmonic (p,q) form. We shall present a proof of these results in 3.1.

We also have

Thm 1.5.2: Holomorphic forms on a Kähler manifold are harmonic.

More generally, holomorphic maps between Kähler manifolds are harmonic.

This expresses the compatibility between the Riemannian and the complex structure.

The equation for a harmonic map between Kähler manifolds becomes in local coordinates

(1.5.6)
$$\gamma^{\alpha\bar{\beta}}\left(\frac{\partial^2}{\partial z^\alpha \partial z^{\bar{\beta}}} f^i + \Gamma^i_{jk}\, \frac{\partial f^j}{\partial z^\alpha}\, \frac{\partial f^k}{\partial z^{\bar{\beta}}} \right) = 0 \qquad\qquad \text{for all } i$$

Also

Thm.1.5.3: For the Betti numbers of a compact Kähler manifold we have

$$(i) \qquad b_k(M) \neq 0 \qquad\qquad \text{if } k \text{ even}$$

$$(ii) \qquad b_k(M) \text{ is even} \qquad\qquad \text{if } k \text{ is odd}$$

pf.: (ii) is a consequence of the Hodge decomposition theorem, whereas (i) holds since $\omega^{\frac{k}{2}}$ represents a nontrivial element of $H^{\frac{k}{2},\frac{k}{2}}(M) \subset H^k(M,\mathbb{C})$ as $\frac{1}{m!}\omega^m = *(1)$ is the volume form of M $(m = dim_{\mathbb{C}} M)$.

Let E be a holomorphic vector bundle over the compact Kähler manifold M, $m = \dim_{\mathbb{C}} M$, $n = \text{rank } E = \dim_{\mathbb{C}} V_z$, where V_z is a fiber, and $g = g_{i\bar{j}} \, dz^i \otimes dz^{\bar{j}}$ be the Kähler metric on M.

For a Hermitian metric on E, we let F be the curvature of the corresponding metric complex connection, i.e. the one compatible with both the holomorphic and the Hermitian structure.

The Hermitian Yang-Mills equation is

$$(1.5.7) \qquad \text{tr}_g F \left(= \frac{i}{2}\Lambda(F) \right) = \mu \, \text{Id} \qquad\qquad (\mu \in \mathbb{R})$$

where $\mu \equiv \text{const}$ and Id is the identity homomorphism of E

If in local coordinates F is represented by $F \in \Omega^{1,1}(\text{End } E))$

$$F^{\beta}_{\ \alpha j \bar{k}} \, dz^j \wedge dz^{\bar{k}}$$

then the equation becomes

$$(1.5.8) \qquad g^{j\bar{k}} F^{\beta}_{\ \alpha j \bar{k}} = \mu \delta_{\alpha\beta} \qquad\qquad (\alpha, \beta = 1, \ldots n)$$

or

$$(1.5.9) \qquad h_{\gamma\bar{\beta}} g^{j\bar{k}} F^{\gamma}_{\ \alpha j \bar{k}} = \mu h_{\alpha\bar{\beta}} \qquad\qquad (h = (h_{\alpha\bar{\beta}}))$$

If $E = T'M$, the holomorphic tangent bundle, and if $h = g$, the equation becomes

$$(1.5.10) \qquad R_{\alpha\bar{\beta}} = g^{j\bar{k}} F_{\alpha\bar{\beta} j\bar{k}} = g_{\gamma\bar{\beta}} g^{j\bar{k}} F^{\gamma}_{\ \alpha j \bar{k}} = \mu g_{\alpha\bar{\beta}}$$

where $R_{\alpha\bar\beta}$ is the Ricci tensor. This equation, namely,

$$(1.5.11) \qquad\qquad \mathrm{Ric} = \mu g$$

is the Kähler-Einstein equation, implying that the Ricci tensor is a constant multiple of the metric tensor.

If E satisfies (1.5.7), it is called Hermite-Einstein or Hermitian Yang-Mills.

Let us discuss the relation between the ordinary and the Hermitian Yang-Mills equation. First of all, write a metric complex connection D on E as $D = d + A = \partial_A + \bar\partial_A$. The Kähler identities (1.5.2), (1.5.3) then also hold for D, i.e.

$$(1.5.12) \qquad\qquad [\Lambda, \bar\partial_A] = -i\partial_A^*, \qquad [\Lambda, \partial_A] = i\bar\partial_A^*$$

and

$$(1.5.13) \qquad\qquad [\Lambda, \bar\partial_A^*] = 0 = [\Lambda, \partial_A^*]$$

The Yang-Mills equation for the curvature $F = F_D$ of D is

$$(1.5.14) \qquad\qquad D^*F = 0$$

and the Bianchi identity is

$$(1.5.15) \qquad\qquad DF = 0$$

Since $D = \partial_A + \bar\partial_A$ and F is of type $(1,1)$, hence $\partial_A F$ is of type$(2,1)$, and $\bar\partial_A F$ of type $(1,2)$, so that (1.5.15) implies

$$(1.5.16) \qquad\qquad \partial_A F = 0 = \bar\partial_A F$$

Using (1.5.12) and (1.5.16), the Yang Mills equation then implies

$$0 = D^*F = -i\,(\bar\partial_A - \partial_A)\Lambda F$$

which, again by considerations of type, is equivalent to

$$(1.5.17) \qquad\qquad D(\Lambda F) = 0$$

This means that, as a section of End E, ΛF is constant, and in particular has constant eigenvalues.

Therefore, if D is a solution of the (ordinary) Yang-Mills equations, then it is a direct sum of Hermite-Einstein connections, i.e. solutions of the Hermitian-Yang-Mills equations. Conversely, if D is a solution of the Hermitian Yang-Mills equations, then

$$\Lambda F = \lambda \; \text{Id} \qquad\qquad \text{with } \lambda \equiv \text{const} \; (\lambda \text{ is purely imaginary})$$

and consequently, (1.5.17) holds trivially.

We also have (cf. [D3])

Lemma 1.5.1: If E has a Hermite-Einstein connection with $\mu < 0$, then the only holomorphic section of E is the zero section. If $\mu = 0$, all holomorphic sections are constant.

pf.: We look at the Laplacians acting on sections of E, i.e. on $\Omega^0(E)$,

$$\Delta_A = D^* D \qquad \text{(we have no term } DD^* \text{ since } \Omega^{-1}(E) = 0)$$

$$\Delta'_A = \partial^*_A \partial_A, \qquad \Delta''_A = \bar\partial^*_A \bar\partial_A$$

From the Kähler identities (1.5.12), (1.5.13) we get $(D = \partial_A + \bar\partial_A)$

(1.5.18)
$$\Delta_A = i\Lambda(\bar\partial_A - \partial_A)(\bar\partial_A + \partial_A)$$
$$= i\Lambda \bar\partial_A \partial_A - i\Lambda \partial_A \bar\partial_A = \Delta'_A + \Delta''_A$$

$$\Delta'_A - \Delta''_A = i\Lambda(\bar\partial_A \partial_A + \partial_A \bar\partial_A) = i\Lambda F \qquad \text{since } F \text{ is a (1,1)-form}$$
$$= 2\mu I \qquad\qquad \text{by assumption.}$$

Hence $\Delta''_A = \frac{1}{2}(\Delta_A - 2\mu I)$. For a holomorphic section s, $\Delta''_A s = 0$. Thus $s \equiv \text{const}$, if $\mu \leq 0$, and $s \equiv 0$, if $\mu < 0$

qed.

For a solution of the Hermitian Yang Mills equation, the constant μ is not arbitrary but is determined by invariants of M and E.

Remember that the first Chern class $c_1(E)$ is represented by

$$(1.5.19) \qquad c_1(E) = \frac{i}{2\pi} h^{\alpha\bar{\beta}} F_{\alpha\bar{\beta}} = \frac{i}{2\pi} h^{\alpha\bar{\beta}} F_{\alpha\bar{\beta}j\bar{k}} \, dz^j \wedge dz^{\bar{k}}$$

With $\omega = \frac{i}{2} g_{j\bar{k}} \, dz^j \wedge dz^{\bar{k}}$

$$c_1(E) \wedge *\omega = \frac{1}{(m-1)!} c_1(E) \wedge \omega^{m-1} = \frac{1}{\pi m!} h^{\alpha\bar{\beta}} g^{j\bar{k}} F_{\alpha\bar{\beta}j\bar{k}} \omega^m$$

Hence from (1.5.8)

$$c_1(E) \wedge *\omega = \mu h_{\alpha\bar{\beta}} h^{\alpha\bar{\beta}} \frac{\omega^m}{\pi m!} = \mu \cdot \text{rank } E \frac{\omega^m}{\pi m!}$$

and putting

$$\deg_\omega E = \int_M c_1(E) \wedge *\omega$$

and integrating,

$$(1.5.20) \qquad \mu = \mu(E) = \frac{\pi}{\text{vol } M} \frac{\deg_\omega E}{\text{rank } E}$$

We note that $\deg_\omega E$ depends on the Kähler form ω but is independent of the Hermitian metric chosen on E, since for Hermitian metrics h_1, h_2, the Chern classes $c_1(E, h_1)$ and $c_1(E, h_2)$ are cohomologous (cf. Lemma 1.4.2), i.e. there exists a 1-form η with

$$c_1(E, h_1) - c_1(E, h_2) = d\eta,$$

hence

$$\int_M (c_1(E, h_1) - c_1(E, h_2)) \wedge *\omega = \int_M d\eta \wedge *\omega$$
$$= \int d(\eta \wedge *\omega) \text{ since } *\omega = \frac{\omega^{m-1}}{(m-1)!} \text{ and } d\omega = 0$$
$$= 0$$

Let us also discuss the Kähler-Einstein equation (1.5.11) more thoroughly. First of all, although in the preceding formal discussion, it appeares as a special case of the Hermitian Yang-Mills equation, it is conceptually different from the latter one. Namely, when one attempts to solve the Hermitian Yang-Mills equations, one fixes the Kähler metric g and seeks a bundle metric h so that the equation is satisfied.

If one wants to solve the Kähler-Einstein metric, one has to look for a Kähler metric g solving the equation.

In order to exhibit the meaning of the equation, we first remark that for a Kähler metric $g_{i\bar{j}}\, dz^i dz^{\bar{j}}$ its Ricci tensor $(R_{i\bar{j}})$ represents the first Chern class of M:

$$(1.5.21) \qquad c_1(M) = \frac{i}{2\pi}\, R_{i\bar{j}}\, dz^i \wedge dz^{\bar{j}} \left(= -\frac{i}{2\pi}\, \partial\bar{\partial} \log\, \det(g_{k\bar{l}}) \right)$$

Conversely, it was asked by Calabi whether for any closed $(1,1)$ form $\frac{i}{2\pi} R'_{i\bar{j}}\, dz^i \wedge dz^{\bar{j}}$ cohomologous to $c_1(M)$, one can find a Kähler metric $(g'_{i\bar{j}})$ with Kähler class cohomologous to the one of $(g_{i\bar{j}})$ with Ricci tensor $(R'_{i\bar{j}})$.

Since the Kähler classes of $(g_{i\bar{j}})$ and $(g'_{i\bar{j}})$ are required to be cohomologous,

$$(1.5.22) \qquad g'_{i\bar{j}} = g_{i\bar{j}} + \frac{\partial^2 \varphi}{\partial z^i \partial z^{\bar{j}}}$$

for some global smooth function φ. Of course, we also impose the requirement that $(g'_{i\bar{j}})$ is positive definite.

Since the Ricci classes are cohomologous, there then exists a smooth function f with

$$(1.5.23) \qquad -\frac{i}{2\pi}\, \partial\bar{\partial} \log\, \det\left(g_{i\bar{j}} + \frac{\partial^2 \varphi}{\partial z^i \partial z^{\bar{j}}} \right) + \frac{i}{2\pi}\, \partial\bar{\partial} \log\, \det(g_{i\bar{j}}) = \frac{i}{2\pi}\, \partial\bar{\partial} f$$

Putting $F = f + \gamma$, with a suitable constant γ, it suffices to solve

$$(1.5.24) \qquad \det\left(g_{i\bar{j}} + \frac{\partial^2 \varphi}{\partial z^i \partial z^{\bar{j}}} \right) \det(g_{i\bar{j}})^{-1} = e^{-F}.$$

Integrating (1.5.24), we get

$$\int e^{-F}\, dM = \int e^{-F} \det(g_{i\bar{j}})\, dz^1 \wedge \ldots \wedge dz^m$$

$$= \int \det\left(g_{i\bar{j}} + \frac{\partial^2 \varphi}{\partial z^i \partial z^{\bar{j}}} \right) dz^1 \wedge \ldots \wedge dz^m$$

$$= \text{vol}\, M \qquad \text{(the volume of M depends only on the}$$

$$\text{cohomology class of the Kähler form)}$$

which determines the above constant γ.

Therefore, given $(g_{i\bar{j}})$ and F, one has to solve (1.5.24) with a function φ so that $g'_{i\bar{j}}$ is positive definite. This was achieved by Yau (cf.[Y1], [Y2]) (some results were also obtained by Aubin).

If one seeks a Kähler-Einstein metric cohomologous to a given one, one has to solve

$$(1.5.25) \qquad \partial\bar{\partial} \log \det \left(g_{i\bar{j}} + \frac{\partial^2 \varphi}{\partial z^i \partial z^{\bar{j}}} \right) = -\mu \left(g_{k\bar{l}} + \frac{\partial^2 \varphi}{\partial z^k \partial z^{\bar{l}}} \right) \, dz^k \wedge dz^{\bar{l}}$$

If the first Chern class $c_1(M)$ is positive (negative), one can choose a Kähler metric $(g_{i\bar{j}})$ for which the Kähler class $\frac{i}{2\pi} g_{i\bar{j}} \, dz^i \wedge dz^{\bar{j}}$ represents $(-)c_1(M)$.

Since the Ricci form $-\frac{i}{2\pi} \partial\bar{\partial} \log \det(g_{k\bar{l}})$ is cohomologous to $c_1(M)$, we have in this case

$$(1.5.26) \qquad -\mu g_{i\bar{j}} = \partial\bar{\partial} \log \, \det(g_{k\bar{l}}) - \partial\bar{\partial} F$$

for some smooth F, with $\mu = \pm 1$. In the same way as before, we see from (1.5.25) and (1.5.26) that it suffices to solve

$$(1.5.27) \qquad \det \left(g_{i\bar{j}} + \frac{\partial^2 \varphi}{\partial z^i \partial z^{\bar{j}}} \right) \det(g_{i\bar{j}})^{-1} = e^{-\mu \varphi - F}$$

This was again achived by Yau ([Y1], [Y2]) in case $\mu = -1$ (and also by Aubin, cf. [A]), settling the Calabi conjecture in this case.

The case $\mu = 1$ is not yet completely solved. To date, the strongest results are due to Tian ([T]).

As a consequence of Yau's results, one can find a Kähler-Einstein metric in the cohomology class of $-c_1(M)$ if $c_1(M)$ is negative. If $c_1(M)$ is zero, one can again find a Kähler-Einstein metric, because we can just put $R'_{i\bar{j}} = 0$ in the above consideration and solve (1.5.24) for this case. Of course, this is also equivalent to (1.5.27) with $\mu = 0$. We obtain a metric with vanishing Ricci curvature.

1.6. The Yang-Mills equation in four dimensions

Let M be a compact oriented Riemannian manifold of real dimension four. Then

$$* : \Lambda^2 T^* M \to \Lambda^2 T^* M,$$

i.e. $\Lambda^2 T^* M$ is mapped into itself. Hence, as $*$ is selfadjoint,

$$\Lambda^2 T^* M = \Lambda^+ \oplus \Lambda^-$$

where $\Lambda^\pm$ is the eigenspace of $*$ w.r.t. the eigenvalue ± 1, and $\oplus$ is L^2-orthogonal sum. $\Lambda^2 T^* M$ has rank 6, and Λ^+ and Λ^- each have rank 3. Λ^+ is called the space of selfdual, Λ^- the space of antiselfdual forms.

If $\alpha \in \Lambda^\pm$, then by definition

(1.6.1) $$\alpha = \pm * \alpha$$

Since

$$|\alpha|^2 * 1 = \alpha \wedge *\alpha,$$

we see

(1.6.2) $$\alpha \in \Lambda^+ \iff \alpha \wedge \alpha = |\alpha|^2 * 1$$

and

(1.6.3) $$\alpha \in \Lambda^- \iff \alpha \wedge \alpha = -|\alpha|^2 * 1$$

If M is Kähler, then Λ^+ consists of the $(2,0)$ and $(0,2)$ forms and the span of the Kähler form ω whereas $\Lambda^- = \text{Ker } \Lambda \cap \{(1,1) - \text{forms}\}$, where Λ was defined in $(1.5.1)$, which is of course orthogonal to ω. This is easily checked by choosing normal coordinates at a given point. Then

$$dz^1 \wedge dz^2, dz^{\bar{1}} \wedge dz^{\bar{2}}, \text{ and } dz^1 \wedge dz^{\bar{1}} + dz^2 \wedge dz^{\bar{2}}$$

satisfy $(1.6.2)$, whereas

$$dz^1 \wedge dz^{\bar{2}}, dz^{\bar{1}} \wedge dz^2, \text{ and } dz^1 \wedge dz^{\bar{1}} - dz^2 \wedge dz^{\bar{2}}$$

satisfy $(1.6.3)$. (Note that $*1 = -\frac{1}{4} dz^1 \wedge dz^{\bar{1}} \wedge dz^2 \wedge dz^{\bar{2}}$.)

Let now a $U(n)$-vector bundle E over the Riemannian manifold M, $\dim_R M = 4$, be given. $n = \operatorname{rank} E = $ fiber dimension of E. Let D be a unitary connection on E, and $F = D^2$ its curvature. Then, cf. (1.4.12),

$$(1.6.4) \qquad c_1(E) = \frac{i}{2\pi} \operatorname{tr} F$$

$$(1.6.5) \qquad c_2(E) - \frac{n-1}{2n} c_1(E) \wedge c_1(E) = \frac{1}{8\pi^2} \operatorname{tr}(F^0 \wedge F^0),$$

where $F^0 = F - \frac{1}{n} \operatorname{tr} F \cdot \operatorname{id}_E$ represents the tracefree part of F. We decompose F^0 into its selfdual and antiselfdual components:

$$F^0 = F^0_+ + F^0_-.$$

Then

$$(1.6.6) \qquad \operatorname{tr}(F^0 \wedge F^0) = \operatorname{tr}(F^0_+ \wedge F^0_+) + \operatorname{tr}(F^0_- \wedge F^0_-)$$

$$\text{since } \Lambda^+ \text{ and } \Lambda^- \text{ are orthogonal}$$

$$= \operatorname{tr}(F^0_+ \wedge *F^0_+) - \operatorname{tr}(F^0_- \wedge *F^0_-)$$

$$= -|F^0_+|^2 + |F^0_-|^2 \qquad \qquad \text{by (1.2.22)}$$

Thus

$$(1.6.7) \qquad (c_2(E) - \frac{n-1}{n} c_1(E)^2)[M] = -\frac{1}{8\pi^2} \int_M (|F^0_+|^2 - |F^0_-|^2) \, dM$$

The Yang-Mills functional decomposes as

$$(1.6.8) \qquad YM(D) = \int_M (|\operatorname{tr} F|^2 + |F^0|^2) \, dM$$

$$= \int_M (|\operatorname{tr} F|^2 + |F^0_+|^2 + |F^0_-|^2) \, dM$$

Since the cohomology class of $\operatorname{tr} F$ is fixed as it represents $c_1(E)$ (1.6.4),

$$\int_M |\operatorname{tr} F|^2$$

is minimal if tr F represents a harmonic 2-form in the class of $c_1(E)$. We can minimize $\int |\text{tr } F|^2$ and $\int |F^0|^2$ independently, and because of the constraint (1.6.7), $\int |F^0|^2$ becomes minimal if

$$F^0_+ = 0 \quad \text{or} \quad F^0_- = 0,$$

depending on the sign of $\left(c_2(E) - \frac{n-1}{2n} c_1(E)^2\right)[M]$, i.e. if

(1.6.9) $$F^0 = \pm * F^0$$

If we have a solution of the Hermitian Yang-Mills equation,

(1.6.10) $$\Lambda F = \lambda \text{Id}_E \qquad (\lambda \text{ purely imaginary}),$$

then

$$F^0 \in \text{Ker } \Lambda.$$

Hence, if F is of type $(1,1)$, then F^0 is antiselfdual, i.e. $F^0 = - * F^0$. Such a connection is called projectively antiselfdual, because F^0 represents the curvature of the associated projective unitary connection.
F itself is not antiselfdual, unless $\lambda = 0$.

If D is a $SU(n)$-connection, then automatically

$$\text{tr } F = 0,$$

since $F \in \Omega^2(\text{Ad } E)$, and the fiber of $\text{Ad } E$ in this case is $\overset{\vee}{u}(n)$ which is tracefree. Thus also, cf. (1.6.4),

$$c_1(E) = 0,$$

and

$$c_2(E)[M] = -\frac{1}{8\pi^2} \int_M \left(|F_+|^2 - |F_-|^2\right) \, dM$$

and

$$YM(D) = \int_M \left(|F_+|^2 + |F_-|^2\right) \, dM$$

so that now the minimum of $YM(D)$ is achieved if and only if F is selfdual or antiselfdual, again depending on the sign of $c_2(E)[M]$. The corresponding equation

$$F = \pm * F$$

is a first order equation for D, and implies the second order equation

$$D^* F = 0$$

by our above reasoning.

We also note that, for $\dim_R M = 4$, the Yang-Mills functional and hence also the Yang-Mills equation is invariant under conformal changes of the metric of M, as we integrate the squared norm of a 2-form w.r.t. the volume form of M.

Finally, let us mention the important result of Uhlenbeck ([U1]) that, for $\dim_R M = 4$, Yang-Mills connections with finite L^2-norm of the curvature cannot have isolated singularities:

<u>Thm. 1.6.1:</u> Let D be a connection in a bundle E over $B^4(0,1)\backslash\{0\}$ and a solution of the Yang-Mills equations in $B^4(0,1)\backslash\{0\}$, where (for simplicity) $B^4(0,1)$ is the unit ball in $\mathbb{R}^4$, with

$$\int_{B^4(0,1)} |F_D|^2 < \infty,$$

$D = d + A$, $A \in H^{1,2}_{loc}(B^4(0,1)\backslash\{0\})$ (i.e. $\varphi A \in H^{1,2}(B^4(0,1))$ for all $\varphi \in C_0^\infty(B^4(0,1)\backslash\{0\}))$. Then D is **gauge equivalent** to a connection $\tilde{D}$ which extends smoothly across 0 to a smooth connection. In particular, the bundle E extends to a bundle on $B^4(0,1)$.

Consequently, by conformal invariance, a Yang-Mills connection on $\mathbb{R}^4$ with finite Yang-Mills norm extends to a smooth Yang-Mills connection on S^4.

cf. [FU] for a simplified proof.

2. Some principles of analysis

2.1. The continuity method and the heat flow method

Let

$$L_t : B_1 \to B_2 \qquad\qquad , \quad B_1, B_2 \text{ Banachspaces}$$

be a family of (partial differential) operators, depending smoothly on a parameter t; typically $t \in [0,1]$ or $t \in [0, \infty)$, and one knows a solution u_0 for $t = 0$, i.e.

$$L_0 u_0 = 0,$$

and one wants to find a solution u_t

$$L_t u_t = 0$$

for all t, in particular either for $t = 1$ or for $t \to \infty$, and in the latter case one would like to have convergence of u_t as $t \to \infty$. The proof usually consists of two steps; namely one shows that

$$\Sigma := \{t : \exists u_t \text{ with } L_t u_t = 0\}$$

is both open and closed (in $[0,1]$ or $[0,\infty)$, resp.). Since by assumption $0 \in \Sigma$, one concludes that a solution exists for every t.

Openness usually is the easier part. Namely, assume $t_0 \in \Sigma$, i.e.

$$L_{t_0} u_{t_0} = 0 \qquad\qquad \text{for some } u_{t_0} \in B_1$$

One wants to show that for small τ,

(2.1.1)
$$L_{t_0 + \tau} u_{t_0 + \tau} = 0$$

is likewise solvable.

We differentiate (2.1.1) w.r.t. τ at $\tau = 0$, and obtain (putting $\frac{d}{d\tau} u_{t_0 + \tau} \big|_{\tau = 0} =: \delta u$ and $\frac{d}{d\tau} L_{t_0}(u_{t_0} + \tau \delta u)_{|\tau = 0} =: \delta L_{t_0}(u_{t_0})(\delta u)$)

(2.1.2)
$$\delta L_{t_0}(u_{t_0})(\delta u) = -\frac{d}{d\tau} L_{t_0 + \tau}(u_{t_0})_{|\tau = 0} =: f$$

(2.1.2) now is a linear equation for $v := \delta u$, and the implicit function theorem shows that we can find a solution of (2.1.1) for small τ, provided we can find a solution $v \in B_1$

$$(2.1.3) \qquad\qquad \delta L_{t_0}(u_{t_0})v = f$$

for every $f \in B_2$ (and provided the homogeneous equation $\delta L_{t_0}(u_{t_0})v = 0$ has finite dimensional kernel; this latter requirement is always satisfied in applications concerning elliptic or parabolic operators, as δL_{t_0} is just the linearization of the operator L_{t_0}). Note that here u_{t_0} is considered as fixed, and we linearize around u_{t_0}.

In other words, we have to show that $\delta L_{t_0}(u_{t_0}) : B_1 \to B_2$ is surjective which is the same as saying that its adjoint has trivial kernel.

Thus in order to show openness, we only have to deal with a linear equation.

The more difficult step then usually is the closedness of Σ, since in interesting cases L_t is a nonlinear operator. If $t_n \to t_0$, and if

$$(2.1.4) \qquad\qquad L_{t_n} u_{t_n} = 0$$

(i.e. one has a solution for t_n), one needs to show that

$$(2.1.5) \qquad\qquad u_{t_n} \to v \qquad\qquad \text{for some } v \in B_1$$

$$\text{(at least for a subsequence of } (t_n))$$

and that

$$(2.1.6) \qquad\qquad L_{t_0} v = 0.$$

In order to get (2.1.6), one has to specify the convergence "$\to$" in (2.1.5); it has to be strong enough that $u_{t_n} \to v$ and (2.1.4) imply (2.1.6).

In order to achieve this, one needs good a-priori estimates

$$(2.1.7) \qquad\qquad \|u_{t_n}\| \le c \qquad\qquad \text{(independent of } n)$$

in a suitable norm $\|\cdot\|$. This norm should be strong enough to guarantee that a compactness property gives the desired convergence at least for a subsequence of (t_n), provided (2.1.4) holds.

Instead of trying to find a solution of

$$L_1 u = 0 \qquad \text{resp.} \qquad L_\infty u = 0$$

by the continuity method, one can also use a parabolic equation. When one wants to find a solution of

(2.1.8) $$Lu = 0, \qquad u = u(x) \qquad \text{, on some space } M,$$

one chooses some initial values g and consideres the parabolic problem

(2.1.9) $$\frac{\partial}{\partial t} u(x,t) - Lu(x,t) = 0 \qquad \text{on } M \times [0,\infty)$$

(2.1.10) $$u(x,0) = g(x)$$

and tries to establish estimates that guarantee that for $t \to \infty$, a solution of (2.1.9) converges to a solution $u(x) = u(x,\infty)$ of (2.1.8).

Again, one has to show open- and closedness of the set of $t \in [0,\infty)$ for which a solution exists. Again, usually closedness is more difficult, whereas openness can be handled with the help of the implicit function theorem. Usually, one also has, because of uniqueness of the solution of (2.1.9), a semigroup property, namely that $u(x, t+s)$ equals $u^t(x,s)$ where $u^t(x,s)$ is the solution of (2.1.9) with initial values $u^t(x,0) = u(x,t)$. This also implies that for openness it suffices to establish short time existence, i.e. that a solution of (2.1.9) exists at least on some small time interval $[0,\varepsilon)$, for given (smooth) initial values g, where $\varepsilon > 0$ may depend on g.

Let us consider the following example:

<u>Lemma 2.1.1:</u> Let N, M be compact Riemannian manifolds, $g \in C^{2+\alpha}(N,M)$. Look at the problem

(2.1.11) $$\frac{\partial}{\partial t} f(x,t) - \tau(f(x,t)) = 0 \qquad \text{for } x \in N, \ 0 \le t < \infty$$

(2.1.12) $$f(x,0) = g(x)$$

where τ is the harmonic map operator (cf. 1.2.).

Then there exists some $\varepsilon > 0$ with the property that (2.1.11) has a solution $f(x,t)$ with $f(x,0) = g(x)$ for $0 \le t \le \varepsilon$.

Moreover, the set $\{T \in (0,\infty) : (2.1.11), (2.1.12)$ has a solution for $0 \le t \le T\}$ is open.

<u>pf.:</u> For example, the space $C^{2+\alpha}(N,M)$ is a Banach manifold, and its tangent space at f is $C^{2+\alpha}(N, f^{-1}TM)$. One knows how to carry over results like the implicit function theorem to such Banach manifolds. Let us pursue a more elementary approach, however.

We embed M differentiably into some $\mathbb{R}^d$, by Whitney's theorem. We can find a tubular neighbourhood $U(M)$ of M, locally diffeomorphic to $M \times B(0,1)^{1)}$.

On $U(M)$, we choose a metric which locally is a product metric, with factors the given Riemannian metric on M and the Euclidean metric on $B(0,1)$. We then continue this metric smoothly to all of $\mathbb{R}^d$, in such a way that it coincides with the standard Euclidean metric outside some ball containing $U(M)$.

We identify M with $M \times \{0\}$; local expressions for the image metric, however, will now refer to the metric just constructed on $\mathbb{R}^d$.

We write, with these conventions,

$$\Lambda^i(f) := \left(\frac{\partial}{\partial t} f(x,t) - \tau(f(x,t)) \right)^i$$

$$= \frac{\partial}{\partial t} f^i - \frac{1}{\sqrt{\gamma}} \frac{\partial}{\partial x^\alpha} \left(\sqrt{\gamma}\, \gamma^{\alpha\beta} \frac{\partial f^i}{\partial x^\beta} \right) - \gamma^{\alpha\beta} \Gamma^i_{jk} \frac{\partial f^j}{\partial x^\alpha} \frac{\partial f^k}{\partial x^\beta}$$

$i = 1, \ldots, d$, where in addition we have chosen local coordinates on N.

By the theory of linear parabolic equations (discussed in more detail in 2.2 below), we first find a solution φ of the linear equation

---

1) $B(0,1) := \{x \in \mathbb{R}^l : |x| \le 1\}, \quad l := d - \dim_\mathbb{R} M$

$$\frac{\partial}{\partial t}\varphi^i - \frac{1}{\sqrt{\gamma}}\frac{\partial}{\partial x^\alpha}\left(\sqrt{\gamma}\,\gamma^{\alpha\beta}\frac{\partial\varphi^i}{\partial x^\beta}\right) - \gamma^{\alpha\beta}\,\Gamma^i_{jk}(g(x))\frac{\partial\varphi^j}{\partial x^\alpha}\frac{\partial g^k}{\partial x^\beta} = 0$$

$$\varphi(x,0) = g(x) \qquad (\in M \subset \mathbb{R}^d)$$

(g is considered here as a fixed given map)

In the sequel, we shall restrict every map to the time interval $[0,\varepsilon)$, for suitable $\varepsilon > 0$.

We now consider the linearization of Λ at $\varphi(\cdot,t)$ (for each fixed t); it is given by

$$\Lambda^i_{\varphi(\cdot,t)}(\psi) = \frac{\partial\psi^i}{\partial t} - \frac{1}{\sqrt{\gamma}}\frac{\partial}{\partial x^\alpha}\left(\sqrt{\gamma}\,\gamma^{\alpha\beta}\frac{\partial\psi^i}{\partial x^\beta}\right)$$

$$- \gamma^{\alpha\beta}\,\Gamma^i_{jk,l}(\varphi(x,t))\psi^l\frac{\partial\varphi^j(x,t)}{\partial x^\alpha}\frac{\partial\varphi^k(x,t)}{\partial x^\beta}$$

$$- \gamma^{\alpha\beta}\,\Gamma^i_{jk}(\varphi(x,t))\left(\frac{\partial\psi^j}{\partial x^\alpha}\frac{\partial\varphi^k(x,t)}{\partial x^\beta} + \frac{\partial\varphi^j(x,t)}{\partial x^\alpha}\frac{\partial\psi^k}{\partial x^\beta}\right)$$

$(i = 1,\ldots,d)$

By the theory of linear parabolic equations, we can solve uniquely for ψ of class $C^{2+\alpha}$ in x, $C^{1+\alpha}$ in t

$$\Lambda_{\varphi(\cdot,t)}(\psi) = h(x,t)$$

$$\psi(x,0) = l(x)$$

for given h of class C^α in x and t and l of class $C^{2+\alpha}$ in x, and $\psi \mapsto \left(\Lambda_{\varphi(\cdot,t)}\psi, \psi(\cdot,0)\right)$ is a continuous bijective linear operator between the corresponding mapping spaces.

By the implicit function theorem, the set of all $\Lambda(\varphi + \sigma)$, σ in a neighbourhood of 0 in the space of functions of class $C^{2+\alpha}$ in x, $C^{1+\alpha}$ in t, covers a neighbourhood of $\Lambda(\psi)$ in the space of C^α-functions.

If we choose $\varepsilon > 0$ small enough, then 0 is contained in this neighbourhood, as

$$\Lambda\varphi(x,t) = 0 \qquad\qquad \text{at } t = 0$$

by construction.

This means, that on some small time interval $[0, \varepsilon]$, we can find a solution $f(x, t)$ of

$$\Lambda f(x, t) = 0$$

$$f(x, 0) = g(x) \ .$$

We now want to show that because $g(x) \in M$, also $f(x, t) \in M$ for all t and all $x \in N$.

We put

$$d_M(p) := \text{dist}^2(p, M) \qquad\qquad \text{for } p \in \mathbb{R}^d$$

where the distance is measured w.r.t. the metric constructed above. One checks that as we locally have a product metric on $U(M)$, on $U(M)$ the Hessian of d_M is nonnegative (the Hessian of a function F is defined as

$$D^2 F(u, v) := \langle \nabla_u \text{ grad } F, v \rangle \qquad\qquad \text{for tangent vectors } u, v) \ .$$

On the other hand, by the chain rule

$$\left(\frac{\partial}{\partial t} - \Delta^- \right) d_M(f(x, t))$$

$$= -D^2 d_M(f_{e^\alpha}, f_{e^\alpha}) + \langle (\text{grad } d_M) \circ f, \left(\frac{\partial}{\partial t} - \tau \right) f \rangle$$

This expression consequently is nonpositive as long as $f(x, t) \in U(M)$ (cf. also Lemma 3.2.1 below). On the other hand, $d_M(f(x, 0)) = 0$, since $f(x, 0) = g(x) \in M$, and therefore, by the maximum principle,

$$d_M(f(x, t)) = 0 \ ,$$

i.e. $f(x, t) \in M$ for all x and t.

Openness then follows by choosing $g(x) = f(x, T)$ as initial values whence the preceding argument implies existence on $[T, T + \varepsilon)$.

qed.

2.2 Elliptic and parabolic Schauder theory

We shall usually obtain our estimates in the Hölder spaces $C^{k,\alpha}$, $k = 0, 1, 2, \ldots$, $0 < \alpha < 1$. The relevant regularity results together with a-priori estimates in the linear case are given by Schauder theory

<u>Thm. 2.2.1:</u> Let $x \in B(0,r) := \{y \in I\!\!R^m : |y| \le 1\}$, $a_{ij}(x)$, $b_i(x)$, $d(x) \in C^\alpha(B(0,r))$ $(i,j = 1,\ldots,m)$, and

$$\lambda|\xi|^2 \le a_{ij}(x)\xi^i \xi^j \le \Lambda|\xi|^2$$

for all $x \in B(0,r)$, $\xi \in I\!\!R^m$, with fixed constants $0 < \lambda \le \Lambda < \infty$

Suppose

$$\frac{\partial}{\partial x^i}\left(a_{ij}(x)\frac{\partial}{\partial x^j}f(x)\right) + b_i(x)\frac{\partial}{\partial x^i}f(x) + d(x)f(x) = g(x)$$

If $d(x) \le 0$, then

$$(2.2.1) \qquad \|f\|_{C^{1,\alpha}(B(0,\frac{r}{2}))} \le c\|g\|_{L^\infty(B(0,r))}$$

and

$$(2.2.2) \qquad \|f\|_{C^{2,\alpha}(B(0,\frac{r}{2}))} \le c\|g\|_{C^\alpha(B(0,r))},$$

and in general

$$(2.2.3) \qquad \|f\|_{C^{1,\alpha}(B(0,\frac{r}{2}))} \le c\left(\|g\|_{L^\infty(B(0,r))} + \|f\|_{L^\infty(B(0,\frac{r}{2}))}\right)$$

and

$$(2.2.4) \qquad \|f\|_{C^{2,\alpha}(B(0,\frac{r}{2}))} \le c\left(\|g\|_{C^\alpha(B(0,r))} + \|f\|_{L^\infty(B(0,\frac{r}{2}))}\right)$$

The constants depend on m, α, $\frac{\Lambda}{\lambda}$, and $\|a_{ij}\|_{C^\alpha(B(0,r))}$, $\|b_i\|_{C^\alpha(B(0,r))}$, $\|d\|_{C^\alpha(B(0,r))}$

Likewise, in the parabolic case, if, for $0 \le t \le T$

$$(2.2.5) \qquad \frac{\partial f(x,t)}{\partial t} - \left(\frac{\partial}{\partial x^i}\left(a_{ij}\frac{\partial}{\partial x^j}f(x,t)\right) + b_i\frac{\partial}{\partial x^i}f(x,t) + d\,f(x,t)\right) = g(x,t)$$

then

$$(2.2.6) \quad \|f(\cdot,t)\|_{C^{1,\alpha}(B(0,\frac{r}{2}))} \leq c \left(\sup_{0 \leq t \leq T} \|g(\cdot,t)\|_{L^{\infty}(B(0,r))} + \sup_{0 \leq t \leq T} \|f(\cdot,t)\|_{L^{\infty}(B(0,\frac{r}{2}))} \right)$$

and

$$(2.2.7) \qquad \|f(\cdot,t)\|_{C^{2,\alpha}(B(0,\frac{r}{2}))} + \|\frac{\partial f}{\partial t}(\cdot,t)\|_{C^{\alpha}(B(0,\frac{r}{2}))}$$

$$\leq c \left(\sup_{0 \leq t \leq T} \|g(\cdot,t)\|_{C^{\alpha}(B(0,r))} + \sup_{0 \leq t \leq T} \|f(\cdot,t)\|_{L^{\infty}(B(0,\frac{r}{2}))} \right)$$

where the constants depend on the same quantities as before.

Remark: The same is true in the vector valued case, i.e. where f, g, a_{ij}, b_i, d take values
in $\mathbb{R}^n$ (instead of $\mathbb{R}$).

These results can also be interpreted as regularity results for weak solutions of elliptic or
parabolic equations. For example if $f \in H^{1,2}(B(0,r))$ satisfies

$$(2.2.8) \int_{B(0,r)} \left(-a_{ij}(x) \frac{\partial}{\partial x^j} f(x) \frac{\partial}{\partial x^i} \varphi(x) + b_i(x) \frac{\partial}{\partial x^i} f(x) \varphi(x) + d(x) f(x) \varphi(x) \right) dx$$

$$= \int_{B(0,r)} g(x) \varphi(x) dx$$

for all $\varphi \in C_0^{\infty}(B(0,r))$,

then f is of class $C^{1,\alpha}$ if g is in L^{∞}, and of class $C^{2,\alpha}$ if g is in C^{α}, together with the
corresponding estimates of the theorem. The reason for the appearance of $\|f\|_{L^{\infty}(B(0,\frac{r}{2}))}$
on the right hand side of (2.2.3), (2.2.4), (2.2.6), and (2.2.7) intuitively is the following:
In the elliptic case, one has to take the possibility of eigenfunctions into account, i.e.
solutions of

$$\frac{\partial}{\partial x^i} \left(a_{ij} \frac{\partial}{\partial x^j} f(x) \right) + \lambda f(x) = 0$$

$\lambda = \text{const} > 0$. In the parabolic case, one also has to take the situation of $f(x,t) = f(t)$
with

$$\frac{\partial}{\partial t} f(x,t) - \frac{\partial}{\partial x^i} \left(a_{ij} \frac{\partial}{\partial x^j} f(x,t) \right) = \frac{\partial}{\partial t} f(t) = c \equiv \text{const}$$

into account. In both these situations, an estimate without the supremum of f on the
right hand side is not valid. One can however obtain a somewhat stronger estimate by
replacing the L^{∞}-norm of f by the L^2-norm of f on the right hand side.

Namely

Thm. 2.2.2: _Suppose on $B(0, r)$, under the assumptions of Thm. 2.2.1_

(2.2.9)
$$\frac{\partial}{\partial x^i}\left(a_{ij}\frac{\partial}{\partial x^j}\,f(x)\right) + b_i(x)\frac{\partial}{\partial x^i}\,f(x) \geq -kf(x),$$

where k is a constant. Then

(2.2.10)
$$\sup_{B(0,\frac{r}{2})}|f| \leq c\frac{1}{r^2}\|f\|_{L^2(B(0,r))},$$

where c depends in the same quantities as before (in particular on k). An analogous result holds in the parabolic case, namely

(2.2.11)
$$\sup_{x \in B(0,\frac{r}{2}),\, T - \frac{\tau}{2} \leq t \leq T}|f(x,t)| \leq c\,r^{-\left(\frac{n+2}{2}\right)}(1 + \tau^{-\frac{1}{2}}r)\|f\|_{L^2(B(0,r) \times \{T - \tau \leq t \leq T\})}$$

(The supremum of $|f(x,t)|$ for $x \in B(0,\frac{r}{2})$, $T - \frac{\tau}{2} \leq t \leq T$ is estimated by the L^2-norm of f on $B(0,r) \times \{T - \tau \leq t \leq T\}$, multiplied by a factor depending on r^{-1} and τ^{-1}, in addition to the various quantities of the data.)

For an account of Schauder theory we refer to [GT], where all the preceding elliptic results can be found. A reference for the parabolic case is [LSU].

2.3. Differential equations on Riemannian manifolds

On a compact Riemannian manifold M, this can be applied as follows.
Suppose

$$(2.3.1) \qquad \Delta_M^- f = g,$$

where Δ_M^- is the Laplace-Beltrami operator, i.e. in local coordinates

$$\Delta_M^- = \frac{1}{\sqrt{g}} \frac{\partial}{\partial x^i} \left(\sqrt{g}\, g^{ij} \frac{\partial}{\partial x^j} \right)$$

We can cover M by balls $B(x_k, \frac{r}{2})$, $r > 0$, $k = 1, \ldots, K$, for which $B(x_k, r)$ is geometrically controlled, i.e. $r \le \min(\frac{\pi}{2\Lambda}, i(M))$, where Λ^2 is a bound for the absolute value of the sectional curvature of M, and $i(M)$ is the injectively radius of M, cf. Lemma 1.3.3. We then get estimates on each ball $B(x_k, \frac{r}{2})$, and since these balls cover M, we get estimates on all of M.

This principle can also be applied if f and g take values in a vector bundle E over M (instead of just in the real numbers). On each ball, E is trivial, $E_{|B(x_k,r)} = B(x_k, r) \times \mathbb{R}^n$. If $e_1, \ldots, e_n$ is a (smooth) basis of sections of E over $B(x_k, r)$ we write

$$f = f^i e_i, \qquad g = g^i e_i$$

If L is a second order elliptic operator on sections of E, we compute on $B(x_k, r)$ from the product rule

$$(2.3.2) \qquad Lf = (Lf^i)e_i + b_{ijk} \frac{\partial f^i}{\partial x^j} e_k + d_{ij} f^i e_j$$

where b_{ijk} involves first derivatives of e_i and d_{ij} is basically $< Le_i, e_j >$, i.e. involves second derivatives of e_i. Hence the equation

$$(2.3.3) \qquad Lf = g$$

becomes the system

$$(2.3.4) \qquad Lf^i + b_{kji} \frac{\partial f^k}{\partial x^j} + d_{ji} f^j = g^i \qquad\qquad (i = 1, \ldots, n),$$

i.e. a system of elliptic equations to which the Theorems 2.2.1 and 2.2.2 apply, cf. the remark following Thm. 2.2.1.

For the Laplace-Beltrami operator Δ_N^- on a Riemannian manifold N, all the preceding estimates actually follow from properties of the approximate fundamental solutions defined in section 1.3. As an example, we provide an elementary proof of results similar to Thm. 2.2.2.

Let us treat the elliptic case first. Thus suppose

$$(2.3.5) \qquad\qquad \Delta^- f \geq -cf$$

where c is a constant. Let us also assume $f \geq 0$.

For simplicity, let us also assume $n = \dim N \geq 3$ (The case $\dim N = 2$ follows by similar arguments).

We put $\frac{n}{2}\rho_0 = \min(i(N), \frac{\pi}{2\Lambda})$, where Λ^2 is a bound for the absolute value of the sectional curvature of N and $i(N)$ is the injectivity radius of N.

For a suitable choice of $\rho \in [\frac{\rho_0}{2}, \rho_0]$, $m \in N$

$$(2.3.6) \qquad\qquad \frac{1}{\rho} \int_{\partial B(m,\rho)} f \leq \frac{c}{\rho^2} \int_{B(m,\rho)} f$$

Then the representation formula (1.3.12) yields, using (2.3.6)

$$(2.3.7) \qquad\qquad f(m) \leq \frac{c}{\rho^2} \int_{B(m,\rho)} \frac{f(x)}{d(x,m)^{n-2}} \, dx$$

We put

$$g_1(m,p) := d(m,p)^{-n+2}$$

$$g_k(m,p) := \int_{d(x,p)\leq\rho} g_{k-1}(m,x) \, g_1(x,p) \, dx$$

We note that

$$(2.3.8) \qquad\qquad g_k(m,p) \leq c \, d(m,p)^{-n+2k}$$

For example, for $k = 2$

$$g_2(m,p) := \int_{d(x,p)\leq\rho} d(m,x)^{-n+2} \, d(x,p)^{-n+2} \, dx$$

W.l.o.g., we can assume $d(m,p) \leq \rho$, and we split the integral into integrals over the regions

$$I := \{x : d(m,x) \leq \frac{1}{2}d(m,p)\}$$

$$II := \{x : d(p,x) \leq \frac{1}{2}d(m,p)\}$$

$$III := B(p,\rho)\backslash(I \cup II)$$

Noting that e.g. on I, $d(p,x) \geq \frac{1}{2}d(m,p)$, we check that

$$g_2(m,p) \leq c\, d(m,p)^{-n+4}$$

as claimed.

In particular, $g_k(m,p)$ is bounded for $2k \geq n$.

We then iterate (2.3.7), i.e. we use (2.3.7) for x instead of m for the integrand $f(x)$ in (2.3.7). After at most $\frac{n}{2}$ steps, by (2.3.8) the integrand of the right hand side becomes bounded, i.e.

$$(2.3.9) \qquad f(m) \leq \frac{c}{\rho^2} \int_{B(m,\frac{n}{2}\rho)} f(x)\, dx$$

Similarly, in the parabolic case, we look at

$$(2.3.10) \qquad \Delta^- f - \frac{\partial}{\partial t}f \geq -cf \qquad\qquad \text{, again assuming } f \geq 0.$$

We shall only need the special case where

$$(2.3.11) \qquad \frac{d}{dt}\int_N f(x)\, dx \leq 0,$$

i.e. the integral of f decreases in time. Thus let us assume (2.3.11).

For $m \in N$, we choose $B(m,\rho_0)$ as before.

We now use the approximate representation formula (1.3.13). Since $\exp(-y) \leq c_\alpha y^{-\alpha}$ for $y > 0$, $\alpha \geq 0$, we obtain

$$(2.3.12) \qquad f(m,t) \leq c_1 \int_{B(m,\rho,t_0,t)} f(x,\tau)(t-\tau)^{-\frac{1}{2}}r(x)^{-n+1}\, dx\, d\tau$$

$$+ \frac{c_n}{\rho^{n+2}} \int_{B(m,\rho,t_0,t)} f + \frac{c_n}{\rho^{n+1}} \int_{\substack{r(x)=\rho \\ t_0 \leq r \leq t}} f$$

$$+ (t-t_0)^{-\frac{n}{2}} \int_{B(m,\rho)} f(x,t_0)\, dx.$$

Here, c_1 depends on n and Λ^2, a bound for the sectional curvature of N.

As before, we can choose $\rho \in [\frac{\rho_0}{2}, \rho]$ with

(2.3.13)
$$\int_{\substack{r(x)=\rho \\ t_0 \leq r \leq t}} f \leq \frac{2}{\rho} \int_{B(m,\rho,t_0,t)} f$$

We define
$$g_1(m,p,t) = t^{-\frac{1}{2}} \cdot d(m,p)^{-n+1}$$
$$g_k(m,p,t) = \int_{\substack{t_0 \leq r \leq t \\ d(x,p) \leq \rho}} g_{k-1}(m,x,t-\tau)\, g_1(x,p,\tau)\, dx\, d\tau$$

and choose $\rho = \rho(p)$ in the definition of g_k in such a way that (2.3.13) is satisfied for p instead of m. We observe that

$$g_k(m,p,t) \leq c_2(t-t_0)^{\frac{k}{2}} d(m,p)^{-n+k}$$

and hence g_k is bounded for $k > n$.

Thus, if we iterate (2.3.12), using (2.3.12) again for $f(x,\tau)$ in the first integral in (2.3.12), we obtain after a finite number of steps

(2.3.14) $f(m,t) \leq c_3 \rho^{-n-2} \displaystyle\int_{B(m,n\rho,t-n(t-t_0),t)} f + c_4(t-t_0)^{-\frac{n}{2}} \int_N f(x,t-n(t-t_0))\, dx.$

In order to locate the last integral at $t - n(t-t_0)$, we have used the fact that $\int f$ decreases in time by (2.3.11).

Choosing $t_0 \geq 0$ in such a way that $t = n(t-t_0)$ and using (2.3.11) again

(2.3.15)
$$f(m,t) \leq c_5\left(t\rho^{-n-2} + t^{-\frac{n}{2}}\right) \int_N f(x,0)\, dx.$$

If we want to avoid the term with $t^{-\frac{n}{2}}$, we can use (1.3.14) instead of (1.3.13) and obtain in a similar way

(2.3.16)
$$f(m,t) \leq c_6 \rho^{-2} \sup_{x \in N} f(x,0).$$

Namely, we then have the term

$$\int_{B(m,\rho)} f(x,0)(t-t_0)^{-\frac{n}{2}} \exp\left(-\frac{r^2(x)}{4(t-t_0)}\right)\, dx$$

which is an approximate solution of the heat equation with initial values $f(x, 0)$, and we use that by the maximum principle the supremum over the space variables of a solution of the heat equation is nonincreasing in time.

We collect these estimates in

Lemma 2.3.1: Suppose f is a solution of (2.3.10) on $[0, t]$ satisfying (2.3.11). If

$$0 < R < \min\left(i(N), \frac{\pi}{2A}\right)$$

(2.3.17)
$$f(m, t) \leq c\left(tR^{-n-2} + t^{-\frac{n}{2}}\right) \int_N f(x, 0) \, dx$$

Furthermore, for any $t_0 < t$, in particular $t_0 = 0$,

(2.3.18)
$$f(m, t) \leq cR^{-2} \sup_{x \in N} f(x, t_0).$$

3. The heat flow on manifolds. Existence and uniqueness of harmonic maps into nonpositively curved image manifolds

3.1. The linear case. Hodge theory by parabolic equations.

In order to warm up, we shall first present the linear case, i.e. look at

(3.1.1)
$$\frac{\partial \alpha(x,t)}{\partial t} - \Delta^- \alpha(x,t) = 0$$

(3.1.2)
$$\alpha(x,0) = \alpha_0(x)$$

where $0 \leq t < \infty$, $x \in N$, where N is a compact Riemannian manifold of dimension n, and $\alpha(\cdot,t)$ and α_0 are k-forms on N.

Of course, this is an equation for a section of $\Lambda^k T^* N$, i.e. our solution α has values in a vector bundle.

Locally, it can therefore be written as a system of linear equations.

The investigation of (3.1.1) and (3.1.2) will lead us to a quick and simple proof of the Hodge theorem, following Milgram-Rosenbloom ([MR]). At the same time, and this is the reason for the inclusion of the present section, we shall encounter several ideas that will be useful in the nonlinear setting of harmonic maps and Yang-Mills equations as well.

Of course, (3.1.1) is a linear parabolic equation, and therefore linear theory implies global existence and uniqueness of a solution for given initial values α_0. For example, one can use Fredholm operator theory or convert (3.1.1) into an integral equation. For reasons of presentation, however, we shall only use short time existence and deduce global existence and uniqueness by elementary arguments which at the same time will suggest the appropriate treatment of nonlinear harmonic map problem.

Short time existence is (cf. Lemma 2.1.1)

Lemma 3.1.1: _Let $\alpha_0 \in \Omega^k$ be of class $C^{2,\alpha}$. Then there exists $\varepsilon > 0$ with the property that (3.1.1) - (3.1.2) has a $C^{2,\alpha}$ solution $\alpha(x,t)$ for $0 \leq t < \varepsilon$._

We now derive decay properties of appropriate integral norms of α as t increases. We first look at

$$(\alpha(\cdot,t),\alpha(\cdot,t)) = \|\alpha(\cdot,t)\|^2 = \int_N \alpha(x,t) \wedge *\alpha(x,t) \qquad \text{(integration w.r.t } x \in N)$$

We compute, for a solution of (3.1.1),

$$(3.1.3) \qquad \frac{d}{dt}\|\alpha(\cdot,t)\|^2 = 2\left(\frac{\partial}{\partial t}\alpha(\cdot,t),\alpha(\cdot,t)\right)$$

$$= 2\left(\Delta^-\alpha(\cdot,t),\alpha(\cdot,t)\right)$$

$$= -2\left(d\alpha(\cdot,t),d\alpha(\cdot,t)\right) - 2\left(d^*\alpha(\cdot,t),d^*\alpha(\cdot,t)\right)$$

$$\text{since } \Delta^- = -(d^*d + dd^*)$$

$$\leq 0$$

We put

$$(3.1.4) \qquad E(\alpha(\cdot,t)) = \frac{1}{2}\left(d\alpha(\cdot,t),d\alpha(\cdot,t)\right) + \frac{1}{2}\left(d^*\alpha(\cdot,t),d^*\alpha(\cdot,t)\right),$$

i.e.

$$\frac{d}{dt}\|\alpha(\cdot,t)\|^2 = -4E(\alpha(\cdot,t))$$

We compute

$$(3.1.5) \qquad \frac{d}{dt}E(\alpha(\cdot,t)) = \left(d\frac{\partial}{\partial t}\alpha(\cdot,t),d\alpha(\cdot,t)\right) + \left(d^*\frac{\partial}{\partial t}\alpha(\cdot,t),d^*\alpha(\cdot,t)\right)$$

$$= -\left(\frac{\partial}{\partial t}\alpha(\cdot,t),\Delta^-\alpha(\cdot,t)\right)$$

$$= -\left(\frac{\partial}{\partial t}\alpha(\cdot,t),\frac{\partial}{\partial t}\alpha(\cdot,t)\right)$$

$$\leq 0$$

In particular

$$(3.1.6) \qquad \frac{d^2}{dt^2}\|\alpha(\cdot,t)\|^2 \geq 0$$

As a consequence of (3.1.3), we obtain the uniqueness of solutions of (3.1.1), (3.1.2).

Lemma 3.1.2: Let $\alpha_1(x,t)$, $\alpha_2(x,t)$ be solutions of (3.1.1) for $0 \le t \le T$

with $\qquad\qquad \alpha_1(x,0) = \alpha_2(x,0)$

Then $\qquad\qquad \alpha_1(x,t) \equiv \alpha_2(x,t) \qquad\qquad$, for $0 \le t \le T$.

pf.: $\alpha(x,t) := \alpha_1(x,t) - \alpha_2(x,t)$ is a solution of (3.1.1) with $\alpha(x,0) = 0$.
(3.1.3) then implies $\alpha(x,t) \equiv 0$ for $0 \le t \le T$.

$\qquad\qquad\qquad\qquad\qquad\qquad\qquad\qquad\qquad\qquad\qquad\qquad\qquad$ qed.

We note that uniqueness implies the following semigroup property of the solutions of (3.1.1):

If $\alpha(\cdot,t)$ solves (3.1.1), then $\alpha(\cdot,t+s) = \alpha_s(\cdot,t)$, where α_s is the solution of (3.1.1) with initial values

$$\alpha_s(\cdot,0) = \alpha(\cdot,s)$$

More generally, we have the following stability result

Lemma 3.1.3: Let $\alpha(x,t,s)$ be a family of solutions of (3.1.1) depending differentiably on a parameter $s \in [0,1]$. Then

$$\frac{d}{dt} \left\| \frac{\partial}{\partial s} \alpha(\cdot,t,s) \right\|^2 \le 0$$

pf.: $\frac{\partial}{\partial s}\alpha(x,t,s)$ is also a solution of (3.1.1), and the claim again follows from (3.1.3).

$\qquad\qquad\qquad\qquad\qquad\qquad\qquad\qquad\qquad\qquad\qquad\qquad\qquad$ qed.

Lemma 3.1.4: Let $\alpha(x,t)$ be a solution of (3.1.1) for $0 \le t < T$ with

$$\alpha(x,0) = \alpha_0(x) \in L^2$$

Then

$$\|\alpha(\cdot,t)\|_{C^{2+\alpha}(N)} + \left\| \frac{\partial}{\partial t} \alpha(\cdot,t) \right\|_{C^{\alpha}(N)} \le c \qquad\qquad \text{, for } 0 < \tau \le t < T,$$

where c depends on $\|\alpha_0\|_{L^2(N)}$, on the geometry of N, and on τ.

pf.: By (3.1.3), $\qquad \|\alpha(\cdot,t)\|_{L^2} \leq \|\alpha_0\|_{L^2}$,

and the claim follows from Thms. 2.2.1 and 2.2.2.

Cor. 3.1.1: Let $\alpha_0 \in C^{2,\beta}$ (for some $\beta \in (0,1)$). Then a solution of

$$\left(\frac{\partial}{\partial t} - \Delta^-\right)\alpha(x,t) = 0$$

$$\alpha(x,0) = \alpha_0(x)$$

exists for all $t \geq 0$.

pf.: If a solution $\alpha(\cdot,t)$ exists for $0 \leq t < T$, then, as $t \to T$, it converges to a $C^{2,\alpha}$ form $\alpha(\cdot,T)$, and local existence (Lemma 3.1.1) implies that the solution can be continued up to some time $T + \varepsilon$.

$\qquad\qquad$ qed.

We now study the behaviour of $\alpha(\cdot,t)$ as $t \to \infty$

Lemma 3.1.5: There exists a sequence (t_n), with $t_n \to \infty$ as $n \to \infty$, for which

$$\frac{\partial \alpha}{\partial t}(x,t_n)$$

converges to zero uniformly in $x \in N$ as $n \to \infty$.

pf.: Since $E(\alpha(\cdot,t)) \geq 0$, (3.1.5) implies that for some sequence $t_n \to \infty$,

$$\left\|\frac{\partial \alpha}{\partial t}(\cdot,t_n)\right\| \to 0$$

Lemma 3.1.4 then implies uniform convergence

$\qquad\qquad$ qed.

Thm. 3.1.1 (Milgram-Rosenbloom): Let $\alpha_0 \in C^{2,\beta}$ be a k-form on N, for some $\beta \in (0,1)$. Then a solution of

$$\left(\frac{\partial}{\partial t} - \Delta^-\right)\alpha(x,t) = 0$$

$$\alpha(x,0) = \alpha_0(x)$$

exists for all $t \in [0,\infty)$, and this solution is unique. As $t \to \infty$, $\alpha(\cdot,t)$ converges in $C^{2,\alpha}$ to a harmonic form $H\alpha$, i.e. $\Delta H\alpha = 0$. If $d\alpha_0 = 0$, then $d\alpha(\cdot,t) = 0$ for all t, and $\int_C \alpha(\cdot,t)$ remains constant for any k-dimensional cycle. In particular, in this case $\int_C H\alpha = \int_C \alpha_0$.

<u>pf.</u>: By Lemmas 3.1.4, 3.1.5, $\alpha(\cdot,t_n)$ converges in $C^{2,\alpha}$ to a harmonic form $H\alpha$, as $t_n \to \infty$.

Putting

$$\alpha_1(x,t) = \alpha(x,t) - H\alpha(x)$$

which is also a solution of (3.1.1), we infer from (3.1.3) and Lemma 3.1.5 that

$$\|\alpha(\cdot,t) - H\alpha(\cdot)\| \to 0$$

as $t \to \infty$, so that $\alpha(\cdot,t)$ converges to $H\alpha$ in L^2 as $t \to \infty$. Lemma 3.1.4 then again implies $C^{2,\alpha}$ convergence.

Uniqueness is contained in Lemma 3.1.2.

If $\alpha(\cdot,t)$ solves (3.1.1), then so does $d\alpha(\cdot,t)$ as $\Delta d = d\Delta$, $d\frac{\partial}{\partial t} = \frac{\partial}{\partial t}d$. Hence if $d\alpha_0 = 0$, then by Lemma 3.1.2 again, $d\alpha(\cdot,t) = 0$ for all t.

Also, if $d\alpha = 0$, $\partial C = 0$,

$$\frac{\partial}{\partial t}\int_C \alpha(\cdot,t) = \int_C \Delta^- \alpha(\cdot,t) = -\int_C d^* d\alpha - \int_{\partial C} d^* \alpha = 0.$$

qed.

Actually, one gets an exponential rate of convergence.

This, can, e.g., be seen as follows:

For fixed $t > 0$, we maximize $\|\alpha(\cdot,t)\|$ for a solution α of (3.1.1) with

$$\|\alpha(\cdot,0)\| = 1, \qquad H\alpha = 0$$

Since, by Thms. 2.2.1, 2.2.2, $|\alpha(\cdot,t)|_{C^{1,\alpha}}$ is bounded in terms of $\|\alpha(\cdot,0)\|$, the maximum is attained; we denote it by $\gamma(t)$. From (3.1.3) $\gamma(t) < 1$. Then

$$\gamma(2t) \leq \gamma(t)^2$$

because of the semigroup property of the solutions of (3.1.1). Likewise

$$\gamma(nt) \leq \gamma(t)^n \qquad \text{for } n \in \mathbb{N}.$$

Consequently

$$\gamma(t) \leq e^{-\lambda t} \qquad \text{for some } \lambda > 0.$$

We easily conclude, for general initial values $\alpha(\cdot, 0) \in L^2$,

(3.1.7) $$\|\alpha(\cdot, t) - H\alpha(\cdot)\| \leq c e^{-\lambda t}$$

If $\beta(x)$ is a k-form on N, then a solution of

(3.1.8) $$\frac{\partial \alpha}{\partial t} - \Delta^- \alpha = \beta$$

(3.1.9) $$\alpha(\cdot, 0) = \alpha_0$$

is obtained in the following way:

For simplicity of notation, if γ is a solution of

$$\frac{\partial \gamma}{\partial t} - \Delta^- \gamma = 0$$

$$\gamma(\cdot, 0) = \alpha_0,$$

we put

$$T_t \alpha_0 := \gamma(\cdot, t)$$

Then, if α solves (3.1.8), (3.1.9),

(3.1.10) $$\alpha(x, t) = T_t \alpha_0(x) + \int_0^t T_{t-\tau} \beta(x) \, d\tau$$

$$= T_t \alpha_0(x) + \int_0^t T_\tau \beta(x) \, d\tau,$$

since β is independent of t.

By (3.1.7),

$$\|T_\tau \beta - H\beta\| \leq e^{-\lambda \tau}$$

Then

$$\|\alpha - tH\beta - T_t \alpha_0\| \leq \int_0^t e^{-\lambda \tau} \, d\tau$$

Thus,

$$\gamma(x) := \lim_{t \to \infty} \left(\alpha(x,t) - tH\beta(x) \right)$$

exists in L^2, and using $C^{2,\alpha}$ estimates as above, also in $C^{2,\alpha}$, and since

$$\left(\frac{\partial}{\partial t} - \Delta^- \right) \left(\alpha(x,t) - tH\beta(x) \right) = \beta(x) - H\beta(x) \qquad \text{, since } \Delta H\beta = 0,$$

we obtain

$$-\Delta^- \gamma = \beta - H\beta$$

and thus

<u>Cor. 3.1.2</u> *(Hodge): Each cohomology class $\alpha_0 \in H^k(N, \mathbb{R})$ (i.e. $\alpha_0 \in \Omega^k(N), d\alpha_0 = 0$) can be represented by a unique harmonic form $H\alpha_0$. A form β is harmonic, i.e. $\Delta^- \beta = 0$, if and only if $d\beta = 0 = d^* \beta$.*

Furthermore, the equation

$$\Delta^- \gamma = \eta$$

is solvable if and only if $(\eta, \beta) = 0$ for all β with $\Delta^- \beta = 0$. γ is then unique up to the addition of a harmonic form. In particular, Ω^k is the L^2 orthogonal sum of the (finite dimensional) kernel of Δ^- and the range of Δ^-.

<u>pf.</u>: All the remaining claims follow by trivial integration by parts arguments; e.g.

$$\Delta^- \beta = 0$$

$$\Rightarrow \qquad 0 = (\Delta^- \beta, \beta) = -(d\beta, d\beta) - (d^* \beta, d^* \beta)$$

$$\Rightarrow \qquad d\beta = 0 = d^* \beta.$$

qed.

Also

<u>Cor. 3.1.3</u> *(Hodge): If N is a Kähler manifold, then*

$$H^k(N, \mathbb{C}) = \bigoplus_{p+q=k} H^{p,q}(N)$$

where $\bigoplus$ is L^2 orthogonal sum

pf.: For a Kähler manifold

$$dd^* + d^* d = 2(\bar\partial\bar\partial^* + \bar\partial^* \bar\partial) = 2(\partial\partial^* + \partial^* \partial),$$

cf. (1.5.4), and consequently, Δ preserves the decomposition

$$\Omega^k = \bigoplus_{p+q=k} \Omega^{p,q},$$

and the claim easily follows from the above, since we get the corresponding orthogonal decomposition for harmonic forms.

qed.

3.2. Harmonic maps

We now look at harmonic maps, i.e. solutions of

$$(3.2.1) \qquad \tau(f(x)) = 0 \qquad (f : N \to M)$$

We want to obtain solutions by looking at the corresponding parabolic problem

$$(3.2.2) \qquad \frac{\partial}{\partial t} f(x,t) = \tau(f(x,t)) \qquad \text{for } t \in [0,\infty)$$

$$(3.2.3) \qquad f(x,0) = g(x)$$

where $g \in C^{2+\alpha}(N,M)$ is given

and show that, as $t \to \infty$, $f(\cdot,t)$ converges to a harmonic map homotopic to g.

We shall always assume for the rest of this chapter that M has <u>nonpositive sectional curvature.</u> It turns out that this assumption implies a certain differential inequality that shall allow us to proceed in a way rather similar to the linear situation of 3.1.

We shall also assume for simplicity that M and N are <u>compact,</u> although in particular the compactness of the image M is not needed in most steps.

We shall also need the following composition property:

If $u \in C^2(N,M)$, $h \in C^2(M,\mathbb{R})$, then

$$\Delta^-(h \circ u) = D^2 h(u_{e^\alpha}, u_{e^\alpha}) + <(\text{grad } h) \circ u, \tau(u) >_M$$

where e^α is an orthonormal frame on N. In particular, if u is harmonic, i.e. $\tau(u) = 0$, this reads as

$$(3.2.4) \qquad \Delta^-(h \circ u) = D^2 h(u_{e^\alpha}, u_{e^\alpha})$$

or in local coordinates

$$\Delta^-(h \circ u) = \gamma^{\alpha\beta} D^2 h(u_{x^\alpha}, u_{x^\beta}).$$

Thus

<u>Lemma 3.2.1:</u> *If h is a (strictly) convex function on M and u is harmonic, then $h \circ u$ is a subharmonic function on N.*

We want to calculate for a harmonic map $f : N \to M$

$$\Delta^- e(f)$$

i.e.

$$\Delta^- \frac{1}{2}\gamma^{\alpha\beta}(x)\, g_{ij}(f(x))\, f^i_{x^\alpha}\, f^j_{x^\beta}.$$

In order to do this, it will be convenient to introduce normal coordinates at the points x and $f(x)$, i.e. $\gamma_{\alpha\beta}(x) = \delta_{\alpha\beta}$ and $g_{ij}(f(x)) = \delta_{ij}$ and all Christoffel symbols vanish at x and $f(x)$, so that we only have to take derivatives of the Christoffel symbols into account which will yield curvature terms eventually.

First of all, we write the equation for harmonic maps in the form

(3.2.5) $$0 = \gamma^{\alpha\beta}\, f^i_{x^\alpha x^\beta} - \gamma^{\alpha\beta}\, {}^N\Gamma^\eta_{\alpha\beta}\, f^i_{x^\eta} + \gamma^{\alpha\beta}\, {}^M\Gamma^i_{kl}\, f^k_{x^\alpha}\, f^l_{x^\beta}.$$

Differentiating this equation at x w.r.t. x^ϵ, we obtain

(3.2.6) $$f^i_{x^\alpha x^\alpha x^\epsilon} = \frac{1}{2}(\gamma_{\alpha\eta,\alpha\epsilon} + \gamma_{\alpha\eta,\alpha\epsilon} - \gamma_{\alpha\alpha,\eta\epsilon})f^i_{x^\eta}$$
$$- \frac{1}{2}(g_{ki,lm} + g_{li,km} - g_{kl,im})f^m_{x^\epsilon} f^k_{x^\alpha} f^l_{x^\alpha},$$

using of course that by our choice of coordinates all first derivatives of the metric tensors vanish, and the Christoffel symbols are given by, e.g. $\Gamma^i_{kl} = \frac{1}{2}g^{im}(g_{mk,l} + g_{ml,k} - g_{kl,m})$. Furthermore, in our coordinates

(3.2.7) $$\gamma^{\alpha\beta}{}_{,\sigma\sigma} = -\gamma_{\alpha\beta,\sigma\sigma}$$

and by the chain rule

(3.2.8) $$\Delta^- g_{ij}(f(x)) = g_{ij,kl} f^k_{x^\sigma} f^l_{x^\sigma}.$$

From (3.2.6) – (3.2.8) we obtain

(3.2.9) $$\Delta^- \frac{1}{2}\gamma^{\alpha\beta}(x)\, g_{ij}(f(x))\, f^i_{x^\alpha} f^j_{x^\beta}$$
$$= f^i_{x^\alpha x^\sigma} f^i_{x^\alpha x^\sigma}$$
$$- \frac{1}{2}(\gamma_{\alpha\beta,\sigma\sigma} + \gamma_{\sigma\sigma,\alpha\beta} - \gamma_{\sigma\alpha,\sigma\beta} - \gamma_{\sigma\alpha,\sigma\beta})f^i_{x^\alpha} f^i_{x^\beta}$$
$$+ \frac{1}{2}(g_{ij,kl} + g_{kl,ij} - g_{ik,jl} - g_{jl,ik})f^i_{x^\alpha} f^j_{x^\alpha} f^k_{x^\sigma} f^l_{x^\sigma}$$
$$= f^i_{x^\alpha x^\sigma} f^i_{x^\alpha x^\sigma} + \frac{1}{2}R^N_{\alpha\beta} f^i_{x^\alpha} f^i_{x^\beta} - \frac{1}{2}R^M_{ikjl} f^i_{x^\alpha} f^j_{x^\alpha} f^k_{x^\sigma} f^l_{x^\sigma},$$

where $R_{\alpha\beta}^N$ is the Ricci tensor of N and R_{ikjl}^M is the curvature tensor of M.

In arbitrary coordinates, this formula is of course transformed into

$$\Delta^- e(f) = g_{ij}(f(x))\,\gamma^{\alpha\beta}(x)\,\gamma^{\sigma\eta}(x)$$
$$\left(f^i_{x^\alpha x^\sigma} - \Gamma^\gamma_{\alpha\sigma}\frac{\partial f^i}{\partial x^\gamma} + \Gamma^i_{kl}\frac{\partial f^k}{\partial x^\alpha}\frac{\partial f^l}{\partial x^\sigma}\right)\left(f^j_{x^\beta x^\eta} - \Gamma^\gamma_{\beta\eta}\frac{\partial f^j}{\partial x^\gamma} + \Gamma^j_{mn}\frac{\partial f^m}{\partial x^\beta}\frac{\partial f^n}{\partial x^\eta}\right)$$
$$+ \frac{1}{2}g_{ij}(f(x))\,R^N_{\alpha\beta}(x)\,f^i_{x^\alpha}f^j_{x^\beta} - \frac{1}{2}\gamma^{\alpha\beta}(x)\,\gamma^{\sigma\eta}(x)\,R^M_{ikjl}(f(x))\,f^i_{x^\alpha}f^j_{x^\beta}f^k_{x^\sigma}f^l_{x^\eta}$$

and in invariant notation, if e_α is an orthonormal frame at x,

(3.2.10)
$$\Delta^- e(f) = |\nabla df|^2 + \frac{1}{2} < df \cdot \mathrm{Ric}^N(e_\alpha), df \cdot e_\alpha >$$
$$- \frac{1}{2} < R^M(df \cdot e_\alpha, df \cdot e_\beta)\, df \cdot e_\beta, df \cdot e_\alpha >$$

In the same way, if $f(x,t)$ is a solution of the parabolic problem (3.2.2), we have

(3.2.11)
$$\Delta^- e(f) - \frac{\partial}{\partial t}e(f) = |\nabla df|^2 + \frac{1}{2} < df \cdot \mathrm{Ric}^N(e_\alpha), df \cdot e_\alpha >$$
$$- \frac{1}{2} < R^M(df \cdot e_\alpha, df \cdot e_\beta)\, df \cdot e_\beta, df \cdot e_\alpha >$$

In the elliptic case, the regularity result is

<u>Thm 3.2.1:</u> If $f : N \to M$ is harmonic, N compact and M nonpositively curved, then

$$|f|_{C^{2+\alpha}(N,M)} \le c$$

where c depends on the energy $E(f)$ and on the geometry of N and M.

pf.: We look at the equation

(3.2.12)
$$\frac{1}{\sqrt{\gamma}}\frac{\partial}{\partial x^\alpha}\left(\sqrt{\gamma}\,\gamma^{\alpha\beta}\frac{\partial}{\partial x^\beta}f^i\right) = -\gamma^{\alpha\beta}\,\Gamma^i_{jk}\frac{\partial f^j}{\partial x^\alpha}\frac{\partial f^k}{\partial x^\beta}$$

Since

$$\Delta e(f) \ge -c\,e(f)$$

since M has nonpositive sectional curvature, cf. (3.2.10), Thm. 2.2.2 gives a pointwise bound on $e(f)$ in terms of $E(f)$.

This implies that the right hand side of (3.2.12) is bounded and that for every $m \in N$, a whole neighbourhood $B(m, \rho)$ is mapped into a single coordinate chart on the image. Elliptic regularity theory implies $f \in C^{1+\alpha}$, which in turn implies that the right hand side is of class C^{α} and hence $f \in C^{2+\alpha}$, cf. Thm. 2.2.1

qed.

3.3 The heat flow for harmonic maps

We first show that the energy $E(f(\cdot, t))$ is a decreasing function of t. For,

$$(3.3.1) \qquad \frac{d}{dt} E(f(\cdot, t)) = \frac{d}{dt} \frac{1}{2} \int <df, df> = \int <\nabla_{\frac{\partial}{\partial t}} df, df> = \int <d\frac{\partial}{\partial t} f, df>$$

$$= -\int <\frac{\partial}{\partial t} f, \tau(f)> = -\int \left|\frac{\partial}{\partial t} f\right|^2$$

since f satisfies the equation (3.2.2), i.e. $\frac{\partial}{\partial t} f = \tau(f)$.

Again, by our curvature assumption, (3.2.11) yields

$$(3.3.2) \qquad \Delta^- e(f) - \frac{\partial}{\partial t} e(f) \geq -c\, e(f).$$

Applying Lemma 2.3.1 to $e(f)$ yields

<u>Lemma 3.3.1:</u> *Let* $t > 0$, $0 < R < \min(i(N), \frac{\pi}{2\lambda})$, *and suppose* f *is a solution of (3.2.2)*
on $[0, t]$. *As always,* M *has nonpositive sectional curvature. Then, for all*
$x \in N$,

$$(3.3.3) \qquad e(f)(x, t) \leq c \left(t R^{-n-2} + t^{-\frac{n}{2}}\right) \int_N e(f)(y, 0)\, dy,$$

and for any $t_0 < t$, *in particular* $t_0 = 0$,

$$(3.3.4) \qquad e(f)(x, t) \leq c\, R^{-2} \sup_{y \in N} e(f)(y, t_0)$$

We now want to derive a stability lemma. In contrast to the linear case, where it was sufficient to consider L^2-norms here we need to consider pointwise quantities.

We let $f(x, t, s)$ be a smooth family of solutions of (3.2.2) depending on a parameter s and having initial values $f(x, 0, s) = g(x, s)$, $0 \leq s \leq s_0$.

<u>Lemma 3.3.2</u> *(Hartman [Ht]) Suppose again, that* M *has nonpositive sectional curvature.*
For every $s \in [0, s_0]$

$$\sup_{x \in N} \left(g_{ij}(f(x, t, s)) \cdot \frac{\partial f^i}{\partial s} \frac{\partial f^j}{\partial s} \right)$$

is nonincreasing in t. Hence also

$$\sup_{x \in N, s \in [0, s_0]} \left(g_{ij} \frac{\partial f^i}{\partial s} \frac{\partial f^j}{\partial s} \right)$$

is a nonincreasing function of t.

pf.: As in 3.2, one calculates in normal coordinates

$$(3.3.5) \qquad \left(\Delta^- - \frac{\partial}{\partial t} \right) \left(g_{ij} \frac{\partial f^i}{\partial s} \frac{\partial f^j}{\partial s} \right) = g_{ij} \frac{\partial^2 f^i}{\partial x^\alpha \partial s} \frac{\partial^2 f^j}{\partial x^\alpha \partial s} - \frac{1}{2} R_{ikjl} \frac{\partial f^i}{\partial s} \frac{\partial f^k}{\partial x^\alpha} \frac{\partial f^j}{\partial s} \frac{\partial f^l}{\partial x^\alpha},$$

and since M has nonpositive sectional curvature, hence

$$(3.3.6) \qquad \left(\Delta^- - \frac{\partial}{\partial t} \right) \left(g_{ij} \frac{\partial f^i}{\partial s} \frac{\partial f^j}{\partial s} \right) \geq 0.$$

The lemma then follows from the maximum principle for parabolic equations.

qed.

We now assume that f_1 and f_2 are smooth homotopic maps from M to N, and $h : N \times [0, 1] \to M$ is a smooth homotopy with $h(x, 0) = f_1(x)$, $h(x, 1) = f_2(x)$.

Since $h(x, s)$ is smooth in x and s, the curve $h(x, \cdot)$ connecting $f_1(x)$ and $f_2(x)$ depends smoothly on x. We let $g(x, \cdot)$ be the geodesic from $f_1(x)$ to $f_2(x)$ which is homotopic to $h(x, \cdot)$ and parametrized proportionally to arc length. Since M is nonpositively curved, this geodesic arc is unique and hence depends smoothly on x. We define $\tilde{d}(f_1(x), f_2(x))$ to be the length of this geodesic arc.

We then put $f(x, 0, s) = g(x, s)$.

<u>Cor. 3.3.1:</u> Suppose, as before, that M is nonpositively curved. Assume that the solution $f(x, t, s)$ of (3.2.2) exists for all $s \in [0, 1]$ and $t \in [0, T]$. Then

$$\sup_{x \in N} \tilde{d}(f(x, t, 0), f(x, t, 1))$$

is nonincreasing in t for $t \in [0, T]$.

pf.: By construction, at $t = 0$

$$\sup_{x \in N, s \in [0,1]} \left(g_{ij} \frac{\partial f^i}{\partial s} \frac{\partial f^j}{\partial s} \right) = \sup_{x \in N} \tilde{d}^2 \left(g(x,0), g(x,1) \right).$$

On the other hand, for any $t \in [0, T]$

$$\tilde{d}^2 \left(f(x,t,0), f(x,t,1) \right) \leq \sup_{s \in [0,1]} g_{ij}(f(x,t,s)) \frac{\partial f^i}{\partial s} \frac{\partial f^j}{\partial s}$$

since $f(x,t,\cdot)$ is a curve joining $f(x,t,0)$ and $f(x,t,1)$ in the homotopy class chosen for the definition of $\tilde{d}$. Using in addition the semigroup property of the heat flow, the claim then follows from Lemma 3.3.2.

qed.

Compared to the linear case, in the Riemannian case we have the problem to make sure that our solutions stay in a given coordinate chart in the image for a fixed amount of time, in order to apply to regularity theorems. Therefore, we have to derive a bound for the time derivative of a solution of (3.2.2) first.

__Lemma 3.3.3:__ *Suppose $f(x,t)$ solves (3.2.2) for $t \in [0,T)$ and M has nonpositive sectional curvature. Then for all $t \in [0,T)$ and $x \in N$*

$$\left| \frac{\partial f(x,t)}{\partial t} \right| \leq \sup_{x \in N} \left| \frac{\partial}{\partial t} f(x,0) \right|.$$

pf.: This follows by putting

$$f(x,t,s) = f(x,t+s)$$

and applying Lemma 3.3.2 at $s = 0$.

qed.

__Lemma 3.3.4:__ *Suppose $f(x,t)$ solves (3.2.2) for $t \in [0,T)$ and M has nonpositive sectional curvature. Then for every $\alpha \in (0,1)$*

(3.3.7)
$$|f(\cdot, t)|_{C^{2+\alpha}(N,M)} + \left| \frac{\partial f}{\partial t}(\cdot, t) \right|_{C^{\alpha}(N,M)} \leq c.$$

c depends on α, the initial values $g(x) = f(x,0)$, and the geometry of N and M.

pf.: We write (3.2.2) in the following way

(3.3.8)
$$\frac{1}{\sqrt{\gamma}}\frac{\partial}{\partial x^\alpha}\left(\sqrt{\gamma}\,\gamma^{\alpha\beta}\,\frac{\partial f^i}{\partial x^\beta}\right) = -\gamma^{\alpha\beta}\,\Gamma^i_{jk}\frac{\partial f^j}{\partial x^\alpha}\frac{\partial f^k}{\partial x^\beta} + \frac{\partial f^i}{\partial t}.$$

If we centre our coordinate charts on N and M at m and $f(m,t_0)$, then for a fixed neighbourhood $B(m,\rho) \times [t_0,t_1]$ of (m,t_0), $f(x,t)$ will stay inside this coordinate chart in M by Lemmata 3.3.1 and 3.3.3. Furthermore, those lemmata also imply that the right hand side of (3.3.8) is bounded. This first implies a bound for $|f(\cdot,t)|_{C^{1+\alpha}(N,M)}$ by elliptic regularity theory, cf. (2.2.3). Then the right hand side of (3.2.2) in local coordinates, i.e. of

(3.3.9)
$$\frac{\partial f^i}{\partial t} - \frac{1}{\sqrt{\gamma}}\frac{\partial}{\partial x^\alpha}\left(\sqrt{\gamma}\,\gamma^{\alpha\beta}\,\frac{\partial f^i}{\partial x^\beta}\right) = \gamma^{\alpha\beta}\,\Gamma^i_{jk}\frac{\partial f^j}{\partial x^\alpha}\frac{\partial f^k}{\partial x^\beta}$$

is bounded in $C^\alpha(N,M)$, and Thm. 2.2.1 then implies the estimate (3.3.7).

qed.

Lemma 3.3.5: The solution of (3.2.2) exists for all $t \in [0,\infty)$, if M has nonpositive sectional curvature.

pf.: Lemma 2.1.1 shows that the set of $T \in [0,\infty)$ with the property that the solution exists for all $t \in [0,T]$ is open and nonempty, while Lemma 3.3.4 implies that it is also closed.

qed.

If we use the energy decay formula (3.3.1), namely

$$\frac{\partial}{\partial t}E(f(\cdot,t)) = -\int_N \left|\frac{\partial f(x,t)}{\partial t}\right|^2\,dN,$$

observe that $E(f(\cdot,t))$ is by definition always nonnegative, and use the time independent C^α-bound for $\left|\frac{\partial f}{\partial t}\right|$, we obtain

<u>Lemma 3.3.6:</u> *There exists a sequence* (t_n), $t_n \to \infty$ *as* $n \to \infty$, *for which* $\frac{\partial f}{\partial t}(x, t_n)$ *converges to zero uniformly in* $x \in N$ *as* $n \to \infty$.

Now using the $C^{2+\alpha}$-bounds for $f(\cdot, t)$ of Lemma 3.3.4, we can assume, by possibly passing to a subsequence, that $f(x, t_n)$ converges uniformly to a harmonic map $f(x)$ as $t_n \to \infty$. In Cor. 3.3.1 which we may apply because of Lemma 3.3.5, we then put

$$g(x, 0) = f(x, 0, 0) = f(x, t_n)$$

$$g(x, s_0) = f(x, 0, s_0) = f(x)$$

By uniform convergence, some $f(\cdot, t_n)$ (and hence all, since $f(x, t)$ is continuous in t) are homotopic to f.

Since $f(x)$ as a harmonic map is a time independent solution of (3.2.2), $f(x, t, s_0) = f(x)$ for all t. Cor. 3.3.1 then implies

$$d(f(x, t_n + t), f(x)) \leq d(f(x, t_n), f(x)) \qquad \text{for all } t \geq 0 .$$

Hence it follows that the selection of the subsequence is not necessary and that $f(x, t)$ uniformly converges to $f(x)$ as $t \to \infty$.

We thus have proved the existence theorem of Eells-Sampson [ES] (with the improvements by Hartman [Ht]) and Al'ber [Al1,2].

<u>Theorem 3.3.1:</u> *Suppose M is nonpositively curved. Then the solution of (3.2.2) exists for all $t \in [0, \infty)$ and converges uniformly to a harmonic map as $t \to \infty$. In particular, any map $g \in C^{2+\alpha}(N, M)$ is homotopic to a harmonic map.*

<u>Cor. 3.3.2:</u> *Any continuous $g : N \to M$ is homotopic to a harmonic map, assuming again that M is nonpositively curved.*

pf.: One just has to smooth out g into a $C^{2,\alpha}$-map. There exist many procedures to do this. For example, cover N by balls $B(x_1, r), \ldots, B(x_k, r)$ with the property that $g(B(x_i, 3r))$ $(i = 1, \ldots, k)$ lies inside a single coordinate chart of M.

Choose $t > 0$, define (in local coordinates) on $B(x_1, 2r)$

$$g_1(x) := \int_0^t \int_{B(x_1, 2r)} \frac{1}{(4\pi\tau)^{\frac{n}{2}}} \exp\left(\frac{-d^2(x,y)}{4\tau}\right) g(y) \, dy \, d\tau,$$

choose smooth functions φ_i with $0 \leq \varphi_i \leq 1$, $\varphi_i \equiv 1$ on $B(x_i, r)$, $\text{supp}(\varphi_i) \subset B(x_i, \frac{3r}{2})$, put

$$\tilde{g}_1 := \varphi_1 g_1 + (1 - \varphi_1) g$$

which is welldefined on the whole of N, put iteratively

$$g_i(x) := \int_0^t \int_{B(x_i, 2r)} \frac{1}{(4\pi r)^{\frac{n}{2}}} \exp\left(\frac{-d^2(x,y)}{4\tau}\right) \tilde{g}_{i-1}(y) \, dy \, d\tau$$

and

$$\tilde{g}_i := \varphi_i g_i + (1 - \varphi_i) \tilde{g}_{i-1}$$

$\tilde{g}_k$ then is smooth, and obviously homotopic to g.

qed.

Remark: Alternatively, one can also show that the heat flow exists for inital values that are only continuous. One only has to check local existence, since after positive time t, the solution becomes automatically smooth. Local existence follows from the appropriate extension of Lemma 2.1.1 to continuous initial values which is a wellknown result in the theory of parabolic equations.

Stronger existence theorems for harmonic maps follow from the work of Schoen-Uhlenbeck [SU]. The existence question for a harmonic map in a given homotopy class does not always have a positive solution, however. For example, Eells-Wood [EW] showed that there is no harmonic map of degree one from a twodimensional torus onto S^2, regardless of the metrics on domain and image. Nevertheless, in two dimensions one knows stronger existence results for harmonic maps than in higher dimensions, cf. e.g. [J1] and [J3].

3.4 Uniqueness of harmonic maps

In this section, we shall be concerned with uniqueness properties of harmonic maps into nonpositively curved manifolds.

Thm. 3.4.1 (Al'ber [Al1,2], Hartman [Ht]) _Let $f_1(x)$, $f_2(x)$ be two homotopic harmonic maps from N into the nonpositively curved manifold M. For fixed x, let $f(x,s)$ be the unique geodesic from $f_1(x)$ to $f_2(x)$ in the homotopy class determined by the homotopy between f_1 and f_2, and let the parameter $s \in [0,1]$ be proportional to arc length._

Then, for every $s \in [0,1]$, $f(\cdot,s)$ is a harmonic map with $E(f(\cdot,s)) = E(f_1) = E(f_2)$. Furthermore, the length of the geodesic $f(x,\cdot)$ is independent of x.

Hence any two homotopic harmonic maps can be joined by a parallel family of harmonic maps with equal energy.

pf.: We let $f(x,t,s)$ be the solution of (3.2.2) with initial values $f(x,0,s) = f(x,s)$. $f(x,t,s)$ exists for all time by Lemma 3.3.5.

By Cor. 3.3.1, for any $s \in [0,1]$ and $t \in (0,\infty)$

(3.4.1)
$$\sup_{x \in N} \bar{d}(f(x,t,s), f_1(x)) \leq \sup_{x \in N} \bar{d}(f(x,s), f_1(x))$$
$$\leq \sup_{x \in N} \bar{d}(f_2(x), f_1(x)).$$

Thm. 3.3.1 implies that $f(x,t,s)$ converges to a harmonic map $f_0(x,s)$ as $t \to \infty$.

We choose $x_0 \in N$ with

$$\bar{d}(f_2(x_0), f_1(x_0)) = \sup_{x \in N} \bar{d}(f_2(x), f_1(x))$$

and by construction therefore

$$\bar{d}(f(x_0,s), f_1(x_0)) = \sup_{x \in N} \bar{d}(f(x,s), f_1(x)) \qquad \text{for all } s \ .$$

From (3.4.1)

(3.4.2)
$$\tilde{d}(f(x_0,t,s), f_1(x_0)) \leq \tilde{d}(f(x_0,s), f_1(x_0))$$

and similarly

(3.4.3)
$$\tilde{d}(f(x_0,t,s), f_2(x_0)) \leq \tilde{d}(f(x_0,s), f_2(x_0)).$$

Note that all distances are measured by the length of that geodesic which is mentioned in the statement of the theorem.

Then (3.4.2) and (3.4.3) imply

(3.4.4)
$$f(x_0,t,s) = f_0(x_0,s) = f(x_0,s) \qquad \text{for all } s.$$

We now look at

$$e_s(f)(x,t,s) = g_{i,j}(f(x,t,s)) \frac{\partial f^i}{\partial s} \frac{\partial f^j}{\partial s}.$$

By Lemma 3.3.2

(3.4.5)
$$\sup_{x \in N} e_s(f)(x,t,s) \leq \sup_{x \in N} e_s(f)(x,0,s) \quad \text{for every } s \in [0,1], \ t \in (0,\infty)$$

On the other hand, from (3.4.4)

(3.4.6)
$$e_s(f)(x_0,t,s) = e_s(f)(x_0,0,s) = \sup_{x \in N} e_s(f)(x,0,s).$$

Hence for all t, the supremum in (3.4.5) is attained at $x = x_0$ and is independent of t. Since by (3.3.6)

(3.4.7)
$$\left(\Delta^- - \frac{\partial}{\partial t} \right) e_s(f) \geq 0,$$

the strong maximum principle implies that $e_s(f)(x,t,s)$ is independent of x and t, i.e.

$$e_s(f)(x,t,s) = e_s(f)(x_0,0,s) \qquad \text{for all } s.$$

Since s is the arc length parameter on the geodesic $f(x_0, \cdot)$, $e_s(f)(x_0,0,s)$ and hence $e_s(f)(x,t,s)$ is also independent of s. Thus for every x and t, $f(x,t,\cdot)$

is a curve of equal length from $f_1(x)$ to $f_2(x)$ parametrized proportionally to arc length. Since $f(x, 0, \cdot)$ was a minimal geodesic, all $f(x, t, \cdot)$ are minimal geodesics and independent of t. In particular $f(x, t, s)$ is time independent for every s, and hence $f(x, 0, s) = f(x, s)$ is harmonic, since $f(x, t, s)$ solves (3.2.2).

Returning to (3.3.6), since $g_{ij} \frac{\partial f^i}{\partial s} \frac{\partial f^j}{\partial s}$ is constant and M is nonpositively curved,

(3.4.8)
$$\frac{\partial^2 f^i}{\partial x^\alpha \partial s} = 0$$

in normal coordinates, or in invariant notation

$$\nabla_{\frac{\partial}{\partial s}} \left(\frac{\partial f^i}{\partial x^\alpha} \frac{\partial}{\partial f^i} \right) = 0$$

where ∇ now is the covariant derivative in the bundle $f^{-1}(x, \cdot) TM$. This implies that the energy density

$$e(f)(x, s) = \gamma^{\alpha\beta}(x) \, g_{ij}(f(x, s)) \frac{\partial f^i}{\partial x^\alpha} \frac{\partial f^j}{\partial x^\beta}$$

is independent of s.

In particular, all the harmonic maps $f(\cdot, s)$ have the same energy.

qed.

Thm. 3.4.2 (Al'ber [Al1,2], Hartman [Ht]) _If M has negative sectional curvature, then a harmonic map $f : N \to M$ is unique in its homotopy class, unless it is constant or maps N onto a closed geodesic. In the latter case, nonuniqueness can only occur by rotations of this geodesic into itself._

pf.: In this case, we see from (3.3.5), that since

(3.4.9)
$$R_{ikjl} \frac{\partial f^i}{\partial s} \frac{\partial f^k}{\partial x^\alpha} \frac{\partial f^j}{\partial s} \frac{\partial f^l}{\partial x^\alpha} = 0$$

by the previous proof, either $\frac{\partial f}{\partial s} \equiv 0$ which means that the family $f(\cdot, x)$ is constant in s and hence consists of a single member, i.e. the harmonic map is unique, or the image of $T_x N$ under df is a one-dimensional subspace of $T_{f(x)} M$. Furthermore, if the harmonic map is not unique, then $f(x, s)$ for any $x \in N$ is a geodesic arc by the construction of the preceding proof. (3.4.9) implies again that df maps $T_x N$ onto the tangent direction of this geodesic. This easily implies that N is mapped into this geodesic.

We now have to show that this geodesic arc extends to a closed geodesic which is covered by $f(N)$.

Since N is compact, f(N) is closed and hence covers some geodesic arc γ. Suppose this arc has an endpoint $p = f(x)$ for some $x \in N$. We choose $q \in \gamma$ within the injectivitiy radius of p. Then $d^2(q, f(y))$ is a subharmonic function on a suitable neighbourhood of $x \in N$ (by Lemmata 3.2.1 and 1.3.3) and has a local maximum at x which is a contradiction, unless $f(y) \equiv p$ for $y \in N$. Thus, if f is not constant, it has to cover a closed geodesic.

qed.

Different proofs of Thms. 3.4.1 and 3.4.2 were obtained by Schoen-Yau [SY].

One also has

Thm. 3.4.3 _([Hm]) Let X be a compact manifold with boundary ∂X, M a manifold with nonpositively sectional curvature._

Let $f_1, f_2 \in C^2(X, M)$ be harmonic and suppose

$$f_{1|\partial X} = f_{2|\partial X}.$$

Then

$$f_1 = f_2$$

pf.: One can apply the maximum principle as before, and one only has to note that now, because the boundary values are fixed, the possible nonuniqueness in Thm. 3.4.2 cannot occur anymore.

qed.

<u>Remark:</u> Actually, one only needs to assume that f_1, f_2 are C^2 in the interior of X and continuous up to the boundary.

One just looks at $X_\epsilon := \{x \in X : d(x, \partial X) \geq \epsilon\}$, restricts f_1 and f_2 to X_ϵ and lets $\epsilon \to 0$. Then

$$\sup_{x \in \partial X_\epsilon} d(f_1(x), f_2(x)) \to 0 \qquad \text{as } \epsilon \to 0,$$

since f_1, f_2 are continuous, and by a stability argument again

$$\sup_{y \in X_\epsilon} d(f_1(y), f_2(y)) \leq \sup_{x \in \partial X_\epsilon} d(f_1(x), f_2(x))$$

and the claim follows.

4. The parabolic Yang-Mills equation

4.1. The parabolic version of the Yang-Mills equation

The main purpose of this chapter is to give a simplified proof of Donaldson's result ([D3]) on the global existence of the Hermitian Yang-Mills flow.

Let us first derive the equation.

We let $\mathcal{A} = \{D = D_0 + A, \ A \in \Omega^1(\mathrm{Ad}E)\}$ be the affine space of connections on a vector bundle E with structure group G over the compact Riemannian manifold M.

The Yang-Mills functional was defined on $\mathcal{A}$ by

$$YM(A) := \|F_A\|^2 = \int |F_A|^2$$

where F_A is the curvature of the connection $D_0 + A$, and the corresponding Euler-Lagrange equations are (cf. (1.2.24))

$$D_A^* F_A = 0,$$

where D_A^* is the adjoint of $D_A = D_0 + A$.

The corresponding parabolic equation then is the gradient flow for YM, namely

(4.1.1)
$$\frac{\partial A}{\partial t} = -D_A^* F_A,$$

and for solutions we have by (1.2.23), looking at $\frac{d}{dt} YM \left(D_0 + t \frac{\partial A}{\partial t} \right)_{|t=0}$,

(4.1.2)
$$\frac{\partial \|F_A\|^2}{\partial t} = -2\|D_A^* F_A\|^2$$

On the other hand, the Yang-Mills functional was invariant under the action of the group of gauge transformations $\mathcal{G}$; thus if $s(t) \in \mathcal{G}$ with $s(0) = \mathrm{id}$, then

$$\mathrm{tr}(F \cdot F) = \mathrm{tr}\big(s^{-1}(t) \ F \ s(t) \ s^{-1}(t) \ F \ s(t)\big),$$

hence

$$0 = \mathrm{tr}\big(-\dot{s}(0) \ F \cdot F\big) + \mathrm{tr}\big(F \cdot F \ \dot{s}(0)\big)$$
$$= \mathrm{tr}\big(F \cdot [F, \dot{s}(0)]\big),$$

hence

$$0 = \int < F, [F, \dot{s}(0)] >$$

$$= \int < F, D_A D_A \dot{s}(0) >$$

$$= \int < D_A^* F, D_A \dot{s}(0) >$$

$\alpha := \dot{s}(0)$ is a section in $C^\infty (\mathrm{Ad} E)$.

Therefore, if we look more generally at

(4.1.3)
$$\frac{\partial A}{\partial t} = -D_A^* F_A + D_A \, \alpha(t),$$

$\alpha(t) \in C^\infty (\mathrm{Ad}\ E)$, then (4.1.2) still holds. If we look at equation (4.1.3), i.e. require that (4.1.1) is satisfied up to an expression $D_A \alpha$, then we can also conveniently pass to a local representation

$$D = d + A,$$

d being the exterior derivative (note that the meaning of A now has changed).

A gauge transformation $D \rightarrow s^*(t)D$ now corresponds to replacing A by $s^* A = s(t)^{-1} ds(t) + s(t)^{-1} A(t) s(t)$, which leads to

$$\frac{\partial}{\partial t} \big(s^*(t)A(t)\big) = d\dot{s}(0) + A\dot{s}(0) - \dot{s}(0)A + \dot{A} \qquad \text{at } t = 0,$$

and as before

$$d\dot{s}(0) + A\dot{s}(0) - \dot{s}(0)A = D_A \alpha$$

Likewise, for arbitrary t

$$\frac{\partial}{\partial t} \big(s^*(t)A(t)\big) = D_{s^*(t)A} s^{-1}(t)\dot{s}(t) + s^{-1}(t)\dot{A}(t)s(t)$$

$$= D_{s^*(t)A} \, \alpha(t) - D_{s^*(t)A}^* F_{s^*(t)A},$$

putting now $\alpha(t) = s^{-1}(t)\dot{s}(t)$ and assuming that (4.1.1) was satisfied for A. For some given t, we may interpret $s(t)$ as well as a coordinate transformation (locally), and we see indeed that a coordinate transformation only amounts to an additional term of the form $D_A \alpha$ in our equation which is irrelevant for the curvature evolution (4.1.2). Therefore, when dealing with (4.1.2), one can safely work with a local representation $D = d + A$.

Lemma 4.1.1: If M is a Kähler manifold and E a Hermitian holomorphic vector bundle on M, then for the curvature of the metric complex connection, we can form ΛF, and with

$$\hat{e} := |\Lambda F|^2 ,$$

we have for a solution of (4.1.3)

$$\left(\frac{\partial}{\partial t} - \Delta^- \right) \hat{e} \leq 0$$

pf.: Since $\hat{e}$ is invariant under the action of $\mathcal{G}$, we may assume w.l.o.g. that

(4.1.4)
$$\frac{\partial A}{\partial t} = -D_A^* F$$

Since

$$F_{D+t\dot{A}} = F_D + t(D\dot{A}) + t^2 \dot{A} \wedge \dot{A},$$

we infer from (4.1.4) that the evolution of the curvature is given by

(4.1.5)
$$\frac{\partial F}{\partial t} = -D_A D_A^* F = -\Delta_A^+ F = \Delta_A^- F,$$

since the Bianchi identity $D_A F = 0$ holds.

We compute

$$\Delta_A^+ \Lambda F - \Lambda \Delta_A^+ F = D_A^* D_A \Lambda F - \Lambda D_A D_A^* F$$

$$\text{(since } D^* = 0 \text{ on zero-forms and } D_A F = 0)$$

$$= (\partial_A^* + \bar{\partial}_A^*)(\partial_A + \bar{\partial}_A)\Lambda F$$
$$\quad - \Lambda(\partial_A + \bar{\partial}_A)(\partial_A^* + \bar{\partial}_A^*)F$$

$$= \Lambda(\partial_A^* + \bar{\partial}_A^*)(\partial_A + \bar{\partial}_A)F + i(\partial_A^* + \bar{\partial}_A^*)(-\bar{\partial}_A^* + \partial_A^*)F$$
$$\quad - (\partial_A + \bar{\partial}_A)(\partial_A^* + \bar{\partial}_A^*)\Lambda F - i(\bar{\partial}_A^* - \partial_A^*)(\partial_A^* + \bar{\partial}_A^*)F$$

$$\text{by } (1.5.12), (1.5.13),$$

$$= 0$$

since $D_A F = 0$, $D_A^* = 0$ on zero-forms, and $\partial_A^* \partial_A^* F = 0 = \bar{\partial}_A^* \bar{\partial}_A^* F$, since F is of type (1,1).

Consequently

$$(4.1.6) \qquad \frac{\partial \Lambda F}{\partial t} - \Delta_A^- \Lambda F = \Lambda \left(\frac{\partial F}{\partial t} - \Delta_A^- F \right) = 0$$

Moreover,

$$\Delta^- \hat{e} = \Delta^- |\Lambda F|^2 = -2 \left(D_A^* D_A \Lambda F, \Lambda F \right) + 2 |D_A \Lambda F|^2,$$

and since on zero-forms,

$$\Delta_A^- = -D_A^* D_A,$$

we obtain from (4.1.6)

$$\frac{\partial \hat{e}}{\partial t} - \Delta^- \hat{e} = -2 |D_A \Lambda F|^2 \leq 0$$

qed.

The parabolic maximum principle then implies

<u>Cor. 4.1.1:</u> *Under the assumptions of Lemma 4.1.1, if $A(x,t)$ is a solution of (4.1.3) for $0 \leq t < T$ with $T \leq \infty$, then*

$$\sup_{x \in M} \hat{e}(x,t) \text{ is nonincreasing in } t \in [0, T),$$

in particular bounded by $\sup_{x \in M} \hat{e}(x,0)$.

For the interpretation of Lemma 4.1.1 and Cor. 4.1.1 note that of course with the connection $A(t)$ also the Hermitian metric on E changes, as discussed in 1.4 and in 4.2 below. Therefore, for each t, $A(t)$ will be the metric complex connection for a different metric.

4.2. The Hermitian Yang-Mills equation and its parabolic analogue

We let again M be a compact Kähler manifold and E a holomorphic vector bundle over M.

Let A_0 be a unitary connection (w.r.t. some Hermitian metric) on E with curvature F_{A_0} of type (1,1).

For $g(t) \in \mathcal{G}^{\sigma}$ ($t \geq 0$), we look at the equation

$$(4.2.1) \qquad g^{-1} \frac{\partial g}{\partial t} = i (\Lambda F_{g \cdot A_0} - \lambda \mathrm{Id}) \qquad (\lambda \text{ purely imaginary})$$

with initial values

$$g(0) = \mathrm{Id}$$

Then the connection $\partial_{g \cdot A_0} + \bar{\partial}_{g \cdot A_0}$ changes via

$$\partial_{(g+t\dot{g}) \cdot A_0} + \bar{\partial}_{(g+t\dot{g}) \cdot A_0} = \overline{(\mathrm{Id} + t g^{-1} \dot{g})}^t \partial_{g \cdot A_0} \overline{(\mathrm{Id} + t g^{-1} \dot{g})}^{t-1}$$
$$+ (\mathrm{Id} + t g^{-1} \dot{g})^{-1} \bar{\partial}_{g \cdot A_0} (\mathrm{Id} + t g^{-1} \dot{g})$$

(putting $\dot{g} = \frac{\partial g}{\partial t}$),

hence for $A(t) = g(t)^* A_0$

$$\frac{\partial A}{\partial t} = -\partial_A \left(\bar{\dot{g}}^t \bar{g}^{t-1} \right) + \bar{\partial}_A \left(g^{-1} \dot{g} \right)$$

Since $\Lambda F \in \Omega^0(\mathfrak{u}(n))$, i.e. $\overline{\Lambda F}^t = -\Lambda F$, we obtain from (4.2.1)

$$\bar{\dot{g}}^t \bar{g}^{t-1} = i (\Lambda F_{g \cdot A_0} - \lambda \mathrm{Id}),$$

hence, again from (4.2.1),

$$\frac{\partial A}{\partial t} = i \bar{\partial}_A \Lambda F - i \partial_A \Lambda F = -D_A^* F \qquad \text{by (1.5.12)},$$

so that (4.2.1) yields a solution of the parabolic Yang-Mills equation.

The action of $\mathcal{G}$, the group of gauge transformations, then corresponds to replacing g by gu with $u\bar{u}^t = \mathrm{Id}$. Then, with $A(t) = (g(t)u(t))^* A_0$,

$$\frac{\partial A}{\partial t} = -\partial_A \left(\left(\bar{\dot{u}}^t \bar{g}^t + \bar{u}^t \bar{\dot{g}}^t \right) \bar{g}^{t-1} \bar{u}^{t-1} \right) + \bar{\partial}_A \left(u^{-1} g^{-1} (g\dot{u} + \dot{g}u) \right)$$
$$= -\partial_A \left(u^{-1} \bar{\dot{g}}^t \bar{g}^{t-1} u \right) + \bar{\partial}_A \left(u^{-1} g^{-1} \dot{g} u \right) - \partial_A \left((u^{-1})^{\cdot} u \right) + \bar{\partial}_A \left(u^{-1} \dot{u} \right)$$

and since $\left(u^{-1}\right)' u = -u^{-1}\dot{u}$,

$$\frac{\partial A}{\partial t} = -D_A^* F + D_A\left(u^{-1}\dot{u}\right)$$

As $u^{-1}\dot{u} \in \Omega^0\left(\text{Ad } E\right)$, we obtain a solution of (4.1.3).

In particular, since $u \in \mathcal{G}$, $g \in \mathcal{G}^\sigma$, each $A(t) = \left(g(t)u(t)\right)^* A_0$ represents a holomorphic structure which is isomorphic to the initial one.

In order to get rid of the action of $\mathcal{G}$, we form $h = g\bar{g}^t$ which is invariant under the action of $\mathcal{G}$. Then, if g is a solution of (4.2.1)

$$\begin{aligned}
\frac{\partial h}{\partial t} &= \frac{\partial g}{\partial t}\bar{g}^t + g\frac{\partial \bar{g}^t}{\partial t} \\
&= g\left(g^{-1}\frac{\partial g}{\partial t}\right)\bar{g}^t + g\left(\frac{\partial \bar{g}^t}{\partial t}\bar{g}^{t^{-1}}\right)\bar{g}^t \\
&= 2i\, g\left(\Lambda F_{g\cdot A_0} - \lambda\text{Id}\right)\bar{g}^t \\
&= 2i\left(\Lambda F_{A_0} + \Lambda\bar{\partial}(h^{-1}\partial h) - \lambda\text{Id}\right)h \qquad\qquad\text{, cf. (1.4.10).}
\end{aligned}$$

Conversely, if we have a solution $h(t)$ of the equation

(4.2.2)
$$\frac{\partial h}{\partial t} = 2i\left(\Lambda F_{A_0} + \Lambda\bar{\partial}(h^{-1}\partial h) - \lambda\text{Id}\right)h$$

with $\overline{h(0)}^t = h(0) = \text{Id}$, then also $\overline{h(t)}^t = h(t)$ for all $t \geq 0$, since $\overline{i(\Lambda F - \lambda\text{Id})}^t = i(\Lambda F - \lambda\text{Id})$ (because $\Lambda F \in \Omega^0\left(\mathscr{K}(n)\right)$ and λ is purely imaginary) and since $\bar{h}^t = h$ implies $\overline{i\partial(h^{-1}\partial h)}^t = i\bar{\partial}(h^{-1}\partial h)$ because the operation of $\bar{\partial}$ and ∂ yields the two-form

$$dz^j \wedge \overline{dz^j} \qquad\qquad \text{(in normal coordinates)}$$
$$= -2i\left(dx^j \wedge dy^j\right) \qquad\qquad \left(z^j = x^j + iy^j\right)$$

Furthermore, if $h(t)$ solves (4.2.2) and $h = g\cdot\bar{g}^t$, then

$$g^{-1}\frac{\partial g}{\partial t} + \frac{\partial \bar{g}^t}{\partial t}\bar{g}^{t^{-1}} = 2i(\Lambda F_{g\cdot A_0} - \lambda\text{Id}),$$

and, with $A(t) = g(t)A_0$,

$$\begin{aligned}
\frac{\partial A(t)}{\partial t} &= i\bar{\partial}_A \Lambda F_A - i\partial_A \Lambda F_A + \frac{1}{2}(\bar{\partial}_A + \partial_A)\left(\frac{\partial \bar{g}^t}{\partial t}\bar{g}^{t^{-1}} - g^{-1}\frac{\partial g}{\partial t}\right) \\
&= -D_A^* F_A + D_A\left(\alpha(t)\right),
\end{aligned}$$

where $\alpha(t) = \frac{1}{2}\left(\dot{g}^t \bar{g}^{t^{-1}} - g^{-1}\dot{g}\right) \in \Omega^0(\mathrm{Ad}\, E)$ $(\mathrm{Ad}\, E = \check{\mathcal{U}}(n)$ here) since $\bar{\alpha}^t = -\alpha$.

Thus, we obtain a solution of (4.1.3).

If we put $H = hH_0$, where H_0 is the initial Hermitian metric on E, then, if (4.2.2) holds

$$(4.2.3) \qquad \frac{\partial H}{\partial t} = 2i\left(\Lambda F_{A_0} + \Lambda\bar{\partial}_{A_0}\left(H_0 H^{-1}\partial_{A_0}(HH_0^{-1})\right) - \lambda\mathrm{Id}\right)H$$

If we linearize this equation at H_0, i.e. insert $H_0 + \varepsilon K$ and differentiate at $\varepsilon = 0$, we obtain

$$(4.2.4) \qquad \frac{\partial K}{\partial t} = 2i\Lambda\bar{\partial}_{A_0}\partial_{A_0}(KH_0^{-1}) + 2i\left(\Lambda F_{A_0} - \lambda\mathrm{Id}\right)K$$
$$= \Delta_{A_0}^-(KH_0^{-1}) + 2i\left(\Lambda F_{A_0} - \lambda\mathrm{Id}\right)K$$

Consequently, if we linearize at a general $H = H(t)$ with corresponding connection A, we just have to replace H_0 by H and A_0 by A, i.e. obtain, putting $\Delta_H := \Delta_A$, etc.,

$$(4.2.5) \qquad \frac{\partial K}{\partial t} = \Delta_H^-(KH^{-1}) + 2i\left(\Lambda F_H - \lambda\mathrm{Id}\right)K$$

This is a (linear) parabolic equation. Hence (4.2.3) is also parabolic, and we obtain short time existence as in Lemma 2.1.1

Lemma 4.2.1: _Given a Hermitian metric H_0 of class $C^{2,\alpha}$ on E, there exists $\varepsilon > 0$ with the property that a solution of (4.2.3) with initial values $H(0) = H_0$ exists for $0 \le t < \varepsilon$ and is of class $C^{2,\alpha}$ w.r.t. $z \in M$, and $\frac{\partial H}{\partial t}$ is of class C^α w.r.t. $z \in M$ $(0 \le t < \varepsilon)$._

Moreover,

$\{T > 0 : a$ _solution with these regularity property exists for_ $0 \le t < T\}$

is open.

For abbreviation, we put

$$\hat{F}_H := \Lambda F_{A_0} + \Lambda\bar{\partial}_{H_0}(h^{-1}\partial_{H_0}h)$$
$$= g\Lambda F_{g \cdot A_0}g^{-1}$$

with $H = hH_0$, $h = g\bar{g}^t$ (e.g. $g = h^{\frac{1}{2}}$).

Then if $K = kH_0 = \eta H$, i.e. $\eta = KH^{-1} = kh^{-1}$,

$$(4.2.6) \qquad \hat{F}_K = \hat{F}_H + \Lambda \bar{\partial}_H (\eta^{-1} \partial_H \eta),$$

cf. (1.4.6).

We also write (4.2.3) as

$$(4.2.7) \qquad \frac{\partial H}{\partial t} = 2i(\hat{F}_H - \lambda \mathrm{Id})H$$

4.3. Global existence

As in 3.3, we shall now exploit certain stability properties of the parabolic Hermitian Yang-Mills equation, initially following [D3].

Def.: Let H, K be Hermitian metrics on E.

$$\tau(H, K) := \text{tr}(K H^{-1})$$

$$\sigma(H, K) := \tau(H, K) + \tau(K, H) - 2 \text{ rank } E$$

We observe that the trace is independent of the choice of a Hermitian base metric H_0. Of course, $\tau(H, K)$ is the sum of the eigenvalues of $H^{-1} K$, and $\tau(K, H)$ the sum of the inverse eigenvalues of $H^{-1} K$. Since, for $\lambda > 0$,

$$\lambda + \frac{1}{\lambda} \geq 2$$

with equality if and only if $\lambda = 1$, we observe

$$\sigma(H, K) \geq 0$$

and

$$\sigma(H, K) = 0 \Longleftrightarrow H = K$$

Lemma 4.3.1: Let $H(t)$, $K(t)$ be solutions of (4.2.7), $\sigma(t) := \sigma(H(t), K(t))$. Then

$$\frac{\partial \sigma}{\partial t} - \Delta^{-} \sigma \leq 0$$

pf.: We shall show that $\tau(t) := \tau(H(t), K(t))$ satisfies

$$\frac{\partial \tau}{\partial t} - \Delta^{-} \tau \leq 0.$$

The equations

$$\frac{\partial H}{\partial t} = 2i(\hat{F}_H - \lambda \text{Id}) H$$

$$\frac{\partial K}{\partial t} = 2i(\hat{F}_K - \lambda \text{Id}) K$$

imply, putting $\eta = KH^{-1}$

$$
\begin{aligned}
\frac{\partial \tau}{\partial t} &= \operatorname{tr}\left(\frac{\partial K}{\partial t}H^{-1} - KH^{-1}\frac{\partial H}{\partial t}H^{-1}\right) \\
&= 2i\,\operatorname{tr}\left(\eta(\hat{F}_K - \hat{F}_H)\right) \\
&= 2i\,\operatorname{tr}\left(\eta\Lambda\bar{\partial}_H\left(\eta^{-1}\partial_H\,\eta\right)\right) && \text{by (4.2.6)} \\
&= \operatorname{tr}\left(\Delta_H^-\eta\right) - 2i\,\operatorname{tr}\left(\Lambda(\overline{\partial_H\eta}^t\,\eta^{-1}\wedge\partial_H\,\eta)\right) && \text{since } \bar{\eta}^t = \eta
\end{aligned}
$$

Since $i\Lambda(\bar{\varphi}^t\wedge\varphi) = |\varphi|^2 \geq 0$, the last term is nonpositive.

Furthermore, since tr is independent of the metric, it commutes with Δ, i.e.

$$
\operatorname{tr}\left(\Delta_H^-\eta\right) = \Delta^-\operatorname{tr}\eta = \Delta^-\tau,
$$

hence

$$
\frac{\partial \tau}{\partial t} - \Delta^-\tau \leq 0
$$

qed.

The maximum principle for parabolic equations yields

<u>Cor. 4.3.1</u>: *A solution of (4.2.7) is uniquely determined by its initial values, i.e. if $H(t)$ and $K(t)$ are solutions of (4.2.7) for $0 \leq t < T$ with $H(0) = K(0)$, then*

$$
H(t) = K(t) \qquad\qquad \text{for } 0 \leq t < T.
$$

We can also use Lemma 4.3.1 to derive a uniqueness result for Hermitian Yang-Mills equations similar to Thm. 3.4.2.

<u>Cor. 4.3.2</u>: *Suppose $H(t)$ is a solution of (4.2.7) for $0 \leq t < T$. Then, as $t \to T$, H_t converges uniformly to some continuous metric H_T.*

pf.: As in 3.3, the stability lemma gives us control in the time direction. Namely, for fixed $\tau > 0$,

$$
\sup_M \sigma(H(t), H(t+\tau))
$$

is a nonincreasing function of t.

Given $\varepsilon > 0$, we find $\delta > 0$ with

$$\sup_M \sigma(H(t), H(t')) < \varepsilon \qquad \text{for } 0 < t, \ t' < \delta,$$

hence

$$\sup_M \sigma(H(t), H(t')) < \varepsilon \qquad \text{for } T - \delta < t, \ t' < T.$$

This shows that $H(t)$ $(T - \delta < t < T)$ satisfies a uniform Cauchy property and hence converges to a nondegenerate limit metric H_T.

qed.

Lemma 4.3.2: Let $H(t)$, $0 \le t < T$, be a family of Hermitian metrics (on the holomorphic bundle E over the compact Kähler manifold M). Suppose that, as $t \to T$, H_t converges uniformly to a metric H_T (not assumed to be smooth), and that $\sup_M |\hat{F}_{H(t)}|$ is bounded independent of $t \in [0, T)$.
Then $H(t)$ is bounded in $C^{1,\alpha}$ independent of $t \in [0, T)$ $(0 < \alpha < 1)$.

pf.: We shall work in local coordinates and define the C^1-norm of $H(t)$ w.r.t. these coordinates.

We suppose that, for some sequence $t_i \to T$,

$$\sup_M |\nabla H(t_i)| = l_i \to \infty$$

Let the supremum be attained at z_i; after selection of a subsequence z_i converges, and we can assume that in our local coordinates, the supremum is attained at 0.

We put

$$\tilde{H}(t_i)(z) = H(t_i)\left(\frac{z}{l_i}\right),$$

hence

$$\left|\nabla \tilde{H}(t_i)\right|(0) = \sup \left|\nabla \tilde{H}(t_i)\right| = 1$$

Then, as $t_i \to T$, $\tilde{H}(t_i)$ converges uniformly to the constant metric $H(T)(0)$

Now

(4.3.1)
$$\hat{F}_{\tilde{H}(t_i)} = \tilde{H}(t_i)^{-1}\left(\Delta\tilde{H}(t_i) - 2i\,\Lambda\bar{\partial}\tilde{H}(t_i)\cdot\tilde{H}(t_i)^{-1}\partial\tilde{H}(t_i)\right)$$

is bounded by assumption, $\tilde{H}(t_i)^{-1}$ is likewise bounded by assumption, since $H(t_i)$ converges to a (nondegenerate) metric, $|\nabla\tilde{H}(t_i)|$ is bounded by construction, hence

(4.3.2)
$$|\Delta\tilde{H}(t_i)(z)| \leq \text{const.}$$

in our coordinate chart, w.l.o.g. for $|z| \leq 1$.

Then, cf. Thm. 2.2.1,

$$|\tilde{H}(t_i)|_{C^{1,\alpha}(\{|z|\leq\frac{1}{2}\})} \leq \text{const.}$$

for each $\alpha \in (0,1)$.

Consequently, after selection of a subsequence, $\tilde{H}(t_i)$ converges strongly in C^1 to a metric $\tilde{H}(T)$.

We conclude $\tilde{H}(T) = H(T)(0)$, hence is constant, contradicting

$$|\nabla\tilde{H}(T)(0)| = \lim_{i\to\infty} |\nabla\tilde{H}(t_i)(0)| = 1.$$

Therefore, the $H(t)$ are bounded in C^1, and

$$\hat{F}_{H(t_i)} = H(t_i)^{-1}\left(\Delta H(t_i) - 2i\,\Lambda\bar{\partial}H(t_i)H(t_i)^{-1}\partial H(t_i)\right)$$

implies as before

$$|H(t)|_{C^{1,\alpha}} \leq \text{const.} \qquad\qquad \text{for each } \alpha \in (0,1).$$

qed.

Thm. 4.3.1: _Let E be a holomorphic vector bundle over the compact Kähler manifold M._
Let λ be purely imaginary. Let H_0 be a Hermitian metric on E of class $C^{2,\alpha}$.
Then

$$\frac{\partial H}{\partial t} = 2i\left(\hat{F}_H - \lambda\mathrm{Id}\right)H$$
$$H(0) = H_0$$

has a unique solution $H(t)$ which exists for $0 \le t < \infty$.

$H(t)$ is of class $C^{2,\alpha}$ w.r.t. $z \in M$ and

$\frac{\partial H}{\partial t}$ is of class C^{α} w.r.t. $z \in M$ for $0 \le t < \infty$.

pf.: Short-time existence was established in Lemma 4.2.1, uniqueness in Cor. 4.3.1. Suppose the solution exists for $0 \le t < T$.

By Cor. 4.1.1, $\sup_{z \in M} |\hat{F}_{H(t)}|$ remains bounded independent of $t \in [0, T)$, and by Cor. 4.3.2, $H(t)$ converges to a continuous metric $H(T)$ as $t \to T$. Hence by Lemma 4.3.2, $H(t)$ is bounded in $C^{1,\alpha}$ independent of $t \in [0, T)$. We then write our equation (4.2.3) as

$$
\begin{aligned}
\frac{\partial H}{\partial t} &= H_0 H^{-1} \Delta_{H_0} (H H_0^{-1}) H \\
&\quad + 2i \, \Lambda \left(\bar{\partial}_{H_0} (H_0 H^{-1}) \wedge \partial (H H_0^{-1}) \right) H \\
&\quad + 2i \left(\Lambda F_{H_0} - \lambda \mathrm{Id} \right) H
\end{aligned}
$$

Since $H(T)$ is a nondegenerate metric, H and H^{-1} are bounded from above and below. By assumption, $H_0 \in C^{2,\alpha}$, hence $\Lambda F_{H_0} \in C^{\alpha}$. Moreover, H is bounded in $C^{1,\alpha}$ independent of $t \in [0, T)$. The regularity theory of linear parabolic equations (Thm. 2.2.1) then implies that H is $C^{2,\alpha}$ w.r.t. $z \in M$ and $\frac{\partial H}{\partial t}$ is C^{α} w.r.t. $z \in M$, with bounds independent of $t \in [0, T)$. Consequently, $H(t) \to H(T)$ in $C^{2,\alpha}$, and short-time existence (Lemma 4.2.1) then implies that the solution continues to exist for $t < T + \varepsilon$, for some $\varepsilon > 0$.

This easily implies the result.

$\qquad$ qed.

In general, however as $t \to \infty$, a solution of

$$
\frac{\partial H}{\partial t} = 2i (\hat{F}_H - \lambda \mathrm{Id}) H
$$

need not converge to a solution of the elliptic equation

$$
\hat{F}_H = \lambda \mathrm{Id}.
$$

This is related to the stability of the bundle E, cf. [Kb], [Lb], [D3], [UY] and the discussion below.

One can only guarantee

Prop. 4.3.1: *Suppose $A(t)$ solves*

$$\frac{\partial A}{\partial t} = -D_A^* F_A + D_A \alpha(t) \qquad\qquad \text{, with } \alpha(t) \in \Omega^0(\text{Ad } E).$$

Then

$$\left(\frac{\partial}{\partial t} - \Delta^-\right) \text{tr } F_A = 0$$

(here, tr is the bundle trace as opposed to the form trace Λ), and consequently, as $t \to \infty$,

$$\text{tr } F_A$$

converges to a harmonic 2-form.

pf.: Since tr F again is invariant under the action of $\mathcal{G}$, it suffices as before to look at

$$\frac{\partial A}{\partial t} = -D_A^* F$$

and the corresponding evolution of the curvature (4.1.5),

$$\frac{\partial F}{\partial t} = \Delta_A^- F$$

Since the trace operator in Ad E is independent of the metric, Δ_A and tr commute, and thus

$$\frac{\partial}{\partial t} \text{tr } F = \Delta^- \text{tr } F.$$

and the claim follows from the result of 3.1.

qed.

Taking up the discussion in 1.5, for a coherent analytic subsheaf $\mathcal{F}$ of a holomorphic vector bundle E over a compact Kähler manifold M, we put

$$\mathcal{F}^* := \text{Hom}(\mathcal{F}, \mathcal{O}) \qquad\qquad \text{, where } \mathcal{O} \text{ is the structure sheaf of } M$$

($\mathcal{F}$ is reflexive if and only if $\mathcal{F}^{**} = (\mathcal{F}^*)^* = \mathcal{F}$; a reflexive sheaf of rank 1 is a holomorphic line bundle, and in general a reflexive sheaf is locally free, i.e. a holomorphic vector bundle, outside a subvariety of codimension at least 2),

$$c_1(\mathcal{F}) := c_1(\det \mathcal{F}^{**})$$
$$\deg_\omega \mathcal{F} := \int_M c_1(\mathcal{F}) \wedge *\omega \qquad (\omega = \text{Kähler form of } M)$$
$$\mu(\mathcal{F}) := \frac{\deg_\omega \mathcal{F}}{\operatorname{rank} \mathcal{F}}$$

(From an analytic point of view, the sheaves $\mathcal{F}$ occuring here can be interpreted as meromorphic sections of $G_M(k,n)$, the bundle over M whose fibers are the Grassmann manifolds $G(k,n)$ of k-dimensional complex subspaces of $\mathbb{C}^n$, $0 < k = \operatorname{rank} \mathcal{F} < n = \operatorname{rank} E$).

The holomorphic vector bundle E is called (semi)stable, if for every coherent subsheaf $\mathcal{F}$ of lower rank,

$$\mu(\mathcal{F}) < \mu(E) \qquad\qquad (\text{resp. } \leq)$$

The result of Uhlenbeck-Yau ([UY]) then is

Thm. 4.3.2: *A stable holomorphic vector bundle over a compact Kähler manifold admits a unique Hermitian Yang-Mills connection.*

A similar result was also obtained by Donaldson ([D4]).

Conversely, the existence of a Hermitian Yang-Mills connection implies stability properties of the bundle, cf. [Kb], [Lb].

5. Geometric applications of harmonic maps

5.1. The topology of Riemannian manifolds of nonpositive sectional curvature

In this section, we shall use our Bochner type formula (3.2.10),

$$(5.1.1) \qquad \Delta^- e(f) = |\nabla df|^2 + \frac{1}{2} < df \cdot \mathrm{Ric}^N(e_\alpha), df \cdot e_\alpha >$$
$$- \frac{1}{2} < R^M (df \cdot e_\alpha, df \cdot e_\beta) \, df \cdot e_\beta, df \cdot e_\alpha >$$

for a harmonic $f : N \to M$ in order to derive some elementary results about the topology of nonpositively curved Riemannian manifolds. These results are well-known, and the present section therefore is included only for reasons of exposition.

<u>Lemma 5.1.1:</u> Let N, M be Riemannian manifolds, N compact, $\mathrm{Ric}^N \geq 0$, $\mathrm{Riem}^M \leq 0^{1)}$,
and let $f : N \to M$ be harmonic.
Then f is totally geodesic, i.e. $\nabla df \equiv 0$, and $e(f) \equiv$ const, and
$< R^M (df e_\alpha, df e_\beta), df e_\beta, df e_\alpha > \equiv 0$
If $\mathrm{Ric}^N(x) > 0$ for some $x \in N$, then $f \equiv$ const.
If $\mathrm{Riem}^M < 0$, then $f(N)$ is a point or a closed geodesic.

<u>pf.:</u>

$$\int_N \Delta e(f) = 0,$$

hence the integral over the right hand side of (5.1.1) vanishes. Since all terms are individually nonnegative, they all have to vanish identically. In particular, $\nabla df \equiv 0$. Also $\Delta e(f) \equiv 0$, hence $e(f) \equiv$ const. If $\mathrm{Ric}^N(x) > 0$, then

$$< df \cdot \mathrm{Ric}^N(e_\alpha), df \cdot e_\alpha > = 0$$

$^{1)}$ this means that all sectional curvatures are nonpositive

implies that at x and hence on all of N, $e(f) = 0$, so that f is constant. Likewise, if $\text{Riem}^M < 0$, then

$$\dim\left(df(T_x N)\right) \le 1 \qquad \text{for all } x \in N,$$

and either this dimension is zero and f is constant, or f as a totally geodesic map has to map N onto a closed geodesic.

Cor. 5.1.1 If M is a compact Riemannian manifold of nonpositive sectional curvature, then

$$\pi_k(M) = 0 \qquad \text{for } k \ge 2,$$

i.e. M is a $K(\pi, 1)$ manifold.

pf.: $\alpha \in \pi_k(M)$ is represented by a map $g : S^k \to M$, where S^k is the standard sphere. By Thm. 3.3.1, g is homotopic to a harmonic map $f : S^k \to M$, and f is constant by Lemma 5.1.1. Hence α is trivial.

qed.

We now want to study the fundamental group of nonpositively curved compact Riemmannian manifolds. The principle idea will be the following:

Let a, b be elements of $\pi_1(M, x_0)$, i.e. have base point x_0. If a and b commute, then the homotopy between ab and ba induces a map $g : T^2 \to M$ where T^2 is a twodimensional torus. We then deform g into a harmonic map $f : T^2 \to M$. During the homotopy between g and f, the base point may change, but of course the two loops corresponding to a and b will always have the same base point at each step of the homotopy.

If now M has negative sectional curvature, then by Lemma 5.1.1, $f(T^2)$ is contained in a closed geodesic c, with base point x_1, say. Then our two loops in $\pi_1(M, x_1)$ are both multiples of c, hence contained in a cyclic subgroup of $\pi_1(M, x_1)$. Then the same is true for the original loops $a, b \in \pi_1(M, x_0)$.

This implies Preissmann's theorem, namely that for a compact negatively curved manifold, every Abelian subgroup of the fundamental group is cyclic. We shall now prove the following generalization of this result due to Gromoll-Wolf [GW] and Lawson-Yau [LY] (cf. also [CE] for a presentation of this and related results):

<u>Thm. 5.1.1:</u> *Let M be a compact Riemannian manifold of nonpositive sectional curvature. Suppose $\Gamma = \pi_1(M)$ contains a solvable subgroup Σ. Then M contains a isometric immersion of a compact flat manifold N with $\pi_1(N) = \Sigma$. Hence, Σ is a Bieberbach group, and in particular finitely generated.*

<u>pf.:</u> First of all, Γ and hence Σ cannot contain elements of finite order. Namely, if $a \in \Gamma$, $ma = 0$ for some $m \geq 2$, we represent a by a closed geodesic, i.e. a harmonic map $f : S^1 \to M$. If $a \neq 0$, then $E(f) \neq 0$.

ma then can be represented by the mth iterate of f which is again harmonic and has energy $mE(f)$. On the other hand, if $ma = 0$, it can also be represented by a constant map, i.e. a map with energy $= 0$. This contradicts Thm. 3.4.1.

Next if Σ is an Abelian group of rank k, then we get a map g from the k-dimensional torus T^k into M. Namely, if a and $b \in \pi_1(M)$ commute, then the homotopy between ab and ba induces a map from T^2 into M. Using that $\pi_k(M) = 0$ for $k \geq 2$ (Cor. 5.1.1), we inductively construct a map from T^k into M. We equip T^k with any flat metric. By Thm. 3.3.1, g is homotopic to a harmonic map $f : T^k \to M$.

By Lemma 5.1.1, $e(f) \equiv$ const, $\nabla df \equiv 0$, and the sectional curvature of M in directions tangent to $f(T^k)$ vanishes. In particular, $f(T^k)$ is flat and $f : T^k \to f(T^k)$ is affine since $\nabla df \equiv 0$. In particular, f has constant rank. If the rank were less than k, then two different generators were mapped onto multiples of the same closed curve, contradicting rank $\Sigma = k$.

Thus, f is an immersion, and by changing the flat structure of T^k, f becomes isometric. Also, $k \leq \dim M$.

Let now Σ be solvable, i.e.

$$\Sigma = \Sigma^n \triangleright \Sigma^{n-1} \triangleright \ldots \triangleright \Sigma^0 = \{e\}$$

where Σ^i is normal in Σ^{i+1} and Σ^{i+1}/Σ^i is Abelian.

In particular, Σ^1 is Abelian, and we have just shown that the claim holds for Σ^1. By induction, we can assume that the result holds for Σ^{n-1}, i.e. that M contains a compact flat manifold N_{n-1} with $\pi_1(N_{n-1}) = \Sigma^{n-1}$.

If now $a \in \Sigma^n \backslash \Sigma^{n-1}$, then since $a\Sigma^{n-1}a^{-1} = \Sigma^{n-1}$, a induces an automorphism of Σ^{n-1}. This in turn induces a homeomorphism of N_{n-1} (since N_{n-1} is a $K(\pi,1)$ manifold because $\pi_i(N_{n-1}) = 0$ for $i \geq 2$; in the present case, the homeomorphism can actually easily be constructed because the universal cover of N_{n-1} is a Euclidean space).

We call this homeomophism g. We then form a twisted product N_n of N_{n-1} and S^1 by identifying in $N_{n-1} \times [0,1]$ $\quad$ $(x,0)$ with $(g(x),1)$. Taking the average of the metrics at $(x,0)$ and $(g(x),1)$, we can easily endow N_n with a flat metric. Since the operation of g on $\pi_i(N_{n-1})$ is conjugation by a, we obtain a map $h: N_n \to M$. Geometrically, this can be visualized as follows: Let $\gamma \in \Sigma^{n-1}$. Then $g_*(\gamma) = a\gamma a^{-1}$. The curves $a\gamma$ and $g_*(\gamma)a$ (considered as elements of $\pi_1(M,x_0)$, for some fixed base point x_0) are homotopic, and as before we get an induced map from the quadrangle with sides γ, a, $g_*(\gamma)^{-1}$, a^{-1} in cyclic order into M, with appropriate boundary identifications. This quadrangle is contained in the boundary of a fundamental region on N_n. Exploiting again $\pi_k(M) = 0$ for $k \geq 2$, we inductively construct a map $h: N_n \to M$ as before.

Then, as before, h is homotopic to a harmonic map $f: N_n \to M$ (Thm. 3.3.1), and f is totally geodesic ($\nabla df \equiv 0$), and the sectional curvature of M vanishes in directions tangent to $f(N_n)$. Also, since a is not homotopic to a curve contained in N_{n-1}, f has maximal rank.

Since Σ^n/Σ^{n-1} is Abelian, we can carry out the preceding construction simultaneously for generators $a_1, \ldots, a_k$ of Σ^n/Σ^{n-1}, i.e. obtain a map from the twisted product N of N_{n-1} and a torus T^k. As before, this map f can be assumed to be harmonic, totally geodesic, and of maximal rank, and again f thus is an affine immersion of N onto its (flat) image. By possibly changing the metric on the torus part of N (the metric on N_{n-1} is already correct by induction), we thus have an isometric immersion of N into M.

qed.

5.2. Siu's strong rigidity theorem for strongly negatively curved Kähler manifolds.

We now want to look at harmonic maps $f : N \to M$ between compact Kähler manifolds, in order to compare two Kähler manifolds that are topologically the same and where we impose a suitable curvature condition on the image manifold M. Of course, the curvature restriction will be some negativity condition on M, and here we have to realize an essential short coming of our formula for $\Delta^- e(f)$, namely that it is of no use if we want to derive any sort of rigidity of the map in case N may have negative Ricci curvature. Of course, since N and M are topologically the same, and M is negatively curved this is precisely what one should expect for N.

If we look at the derivation of the formula for $\Delta e(f)$, we see that the term involving the Ricci curvature of the domain came from the contraction by the inverse metric of the domain.

It was then Siu's idea to avoid this contraction and consequently work with a form instead of a function.

In most of this section, we shall follow [S1]; cf. also [S2], [S3], [S4], [Wu].

Let us fix some notation first.

We now put $m = \dim_{\mathbb{C}} M$, $n = \dim_{\mathbb{C}} N$, i.e. use complex instead of real dimensions. We shall usually assume $n \geq 2$.

$(z^1, \ldots, z^n)$ and $(f^1, \ldots, f^m)$ will be local holomorphic coordinates on N and M, resp., $\gamma_{\alpha \bar{\beta}} \, dz^{\alpha} \, dz^{\bar{\beta}}$ and $g_{i\bar{j}} \, df^i \, df^{\bar{j}}$ the corresponding Kähler metrics. $\omega := i \, \gamma_{\alpha \bar{\beta}} \, dz^{\alpha} \wedge dz^{\bar{\beta}}$ is the Kähler form of N.

$$(5.2.1) \qquad R_{i\bar{j}k\bar{l}} = \frac{\partial}{\partial f^{\bar{l}}} \left(\frac{\partial}{\partial f^k} g_{i\bar{j}} \right) - g^{s\bar{t}} \frac{\partial}{\partial f^k} g_{i\bar{t}} \frac{\partial}{\partial f^{\bar{l}}} g_{s\bar{j}}$$

is the curvature tensor of M. The sectional curvature of the plane spanned by $p = 2 \mathrm{Re} \xi^i \frac{\partial}{\partial f^i}$ and $q = 2 \mathrm{Re} \eta^j \frac{\partial}{\partial f^j}$ is

$$(5.2.2) \qquad \frac{1}{|p \wedge q|^2} R_{i\bar{j}k\bar{l}} \left(\xi^i \eta^{\bar{j}} - \eta^i \xi^{\bar{j}} \right) \left(\xi^k \eta^{\bar{l}} - \eta^k \xi^{\bar{l}} \right)$$

We put, e.g.,

$\bar{\partial} f^i = \frac{\partial f^i}{\partial z^\alpha} dz^\alpha$, and also for short $\partial_\alpha f^i = \frac{\partial f^i}{\partial z^\alpha}$, etc.

$\bar{\partial} f$ then is a section of $(T^{0,1} N)^* \otimes f^* T^{1,0} M$.

As above, the connection $f^* \nabla$ (∇ = Kähler connection on $T^{1,0} M$) can be combined with the exterior derivative on $(T^{0,1} N)^*$ to give, e.g. for the ∂-part

$$D' \omega = \left(\partial_\alpha \omega^i_{\bar{\beta}} + {}^M\Gamma^i_{jk} \, \omega^j_{\bar{\beta}} \, \partial_\alpha f^k \right) \left(dz^\alpha \wedge dz^{\bar{\beta}} \right) \otimes \frac{\partial}{\partial f^i}$$

($\omega = \omega^i_\alpha \, dz^\alpha \otimes \frac{\partial}{\partial f^i}$); cf. chapter 1 for a more general discussion of these constructions.

In particular, we have

$$D' \bar{\partial} f^i = \partial \bar{\partial} f^i + \Gamma^i_{jk} \, \partial f^j \wedge \bar{\partial} f^k$$

We then have Siu's Bochner type identity which holds for any smooth, not necessarily harmonic map $f : N \to M$.

Lemma 5.2.1:

(5.2.3) $\quad \partial \bar{\partial} \left(g_{i\bar{j}} \bar{\partial} f^i \wedge \partial f^{\bar{j}} \right) = R_{i\bar{j}k\bar{l}} \, \bar{\partial} f^i \wedge \partial f^{\bar{j}} \wedge \partial f^k \wedge \overline{\partial f^l} - g_{i\bar{j}} D' \bar{\partial} f^i \wedge D'' \partial f^{\bar{j}}$

pf.: Let $z \in N$ and choose holomorphic normal coordinates at $f(z)$ so that at $f(z)$ all first derivatives of $g_{i\bar{j}}$ vanish.

Then, using the chain rule, at z,

$$\begin{aligned}
\partial \bar{\partial} \left(g_{i\bar{j}}(f) \bar{\partial} f^i \wedge \partial f^{\bar{j}} \right) &= \left(\partial_k \partial_l g_{i\bar{j}} \right) \partial f^k \wedge \bar{\partial} f^l \wedge \bar{\partial} f^i \wedge \partial f^{\bar{j}} \\
&+ \left(\partial_{\bar{k}} \partial_l g_{i\bar{j}} \right) \partial f^{\bar{k}} \wedge \bar{\partial} f^l \wedge \bar{\partial} f^i \wedge \partial f^{\bar{j}} \\
&+ \left(\partial_k \partial_{\bar{l}} g_{i\bar{j}} \right) \partial f^k \wedge \bar{\partial} f^{\bar{l}} \wedge \bar{\partial} f^i \wedge \partial f^{\bar{j}} \\
&+ \left(\partial_{\bar{k}} \partial_{\bar{l}} g_{i\bar{j}} \right) \partial f^{\bar{k}} \wedge \bar{\partial} f^{\bar{l}} \wedge \bar{\partial} f^i \wedge \partial f^{\bar{j}} \\
&- g_{i\bar{j}} \partial \bar{\partial} f^i \wedge \bar{\partial} \partial f^{\bar{j}}
\end{aligned}$$

Now $\partial_k \partial_l g_{i\bar{j}} = \partial_k \partial_i g_{l\bar{j}}$, whereas $\partial f^k \wedge \bar{\partial} f^l \wedge \bar{\partial} f^i \wedge \partial f^{\bar{j}}$
$= -\partial f^k \wedge \bar{\partial} f^i \wedge \bar{\partial} f^l \wedge \partial f^{\bar{j}}$ hence the first term vanishes.

The second and the fourth likewise vanish, while (in our coordinates)

$$\partial_k \partial_{\bar{l}} g_{i\bar{j}} = R_{i\bar{j}k\bar{l}}$$

and

$$\partial \bar{\partial} f^i = D' \bar{\partial} f^i,$$

and the formula follows.

In order to apply this formula, we need to control the terms on the right hand side of (5.2.3).

Lemma 5.2.2: If $f : N \to M$ is harmonic, then

(5.2.4) $$g_{i\bar{j}} D' \bar{\partial} f^i \wedge D'' \partial f^{\bar{j}} \wedge \omega^{n-2} = a \omega^n$$

for some nonpositive function a on N

pf.: We choose holomorphic normal coordinates at z and $f(z)$, and write

$$f^j = u^j + i v^j \qquad (= \operatorname{Re} f^j + i \operatorname{Im} f^j).$$

Then

$$\partial \bar{\partial} f^i \wedge \bar{\partial} \partial f^{\bar{i}} = \partial \bar{\partial} u^i \wedge \bar{\partial} \partial u^i + \partial \bar{\partial} v^i \wedge \bar{\partial} \partial v^i$$

Let $\lambda_1^i, \ldots, \lambda_n^i$ be the eigenvalues of $\left(\dfrac{\partial^2 u^i}{\partial z^\alpha \, \partial z^{\bar{\beta}}} \right)$.

Then, by diagonalizing this Hessian, we see that

$$\partial \bar{\partial} u^i \wedge \bar{\partial} \partial u^i \wedge \omega^{n-2} = \frac{4}{n(n-1)} \sum_i \sum_{\alpha \neq \beta} \lambda_\alpha^i \lambda_\beta^i \omega^n$$

(the factor (-1) from $\bar{\partial} \partial u^i = -\partial \bar{\partial} u^i$ cancels with the factor i^2 from ω^2)

If $\mu_1^i, \ldots, \mu_n^i$ are the eigenvalues of $\left(\dfrac{\partial^2 v^i}{\partial z^\alpha \, \partial z^{\bar{\beta}}} \right)$, we get a similar formula, and hence in our coordinates

$$g_{i\bar{j}} D' \bar{\partial} f^i \wedge D'' \partial f^{\bar{j}} \wedge \omega^{n-2} = \partial \bar{\partial} f^i \wedge \bar{\partial} \partial f^{\bar{i}} \wedge \omega^{n-2}$$

$$= \frac{4}{n(n-1)} \sum_i \sum_{\alpha \neq \beta} \left(\lambda_\alpha^i \lambda_\beta^i + \mu_\alpha^i \mu_\beta^i \right) \omega^n$$

On the other hand, since f is harmonic

$$\sum_\alpha \lambda_\alpha^i = 0,$$

hence

$$\sum_{\alpha \neq \beta} \lambda_\alpha^i \lambda_\beta^i = (\sum_\alpha \lambda_\alpha^i)^2 - \sum_\alpha \lambda_\alpha^{i\,2} = -\sum_\alpha \lambda_\alpha^{i\,2}$$

and the conclusion follows.

qed.

Moreover, if $\gamma_{\alpha\bar\beta} = \delta_{\alpha\beta}$ at z,

$$R_{i\bar j k \bar l}\, \bar\partial f^i \wedge \partial f^{\bar j} \wedge \partial f^k \wedge \overline{\partial f^l} \wedge \omega^{n-2}$$

$$= i^{n-2}(n-2)!\, R_{i\bar j k \bar l}\left(-\partial_\alpha f^i \partial_\alpha f^{\bar j} \partial_\beta f^k \partial_{\bar\beta} f^{\bar l} + \partial_\alpha f^i \partial_\beta f^{\bar j} \partial_\alpha f^k \partial_{\bar\beta} f^{\bar l}\right.$$

$$\left.+\partial_{\bar\beta} f^i \partial_\alpha f^{\bar j} \partial_\beta f^k \partial_\alpha f^{\bar l} - \partial_{\bar\beta} f^i \partial_\beta f^{\bar j} \partial_\alpha f^k \partial_\alpha f^{\bar l}\right)$$

$$dz^1 \wedge dz^{\bar 1}\ldots \wedge dz^n \wedge dz^{\bar n}$$

$$(5.2.5) \qquad = \frac{4}{n(n-1)} R_{i\bar j k \bar l}\left(\partial_\alpha f^i \overline{\partial_\beta f^j} - \partial_{\bar\beta} f^i \overline{\partial_\alpha f^j}\right)\overline{\left(\partial_\alpha f^l \overline{\partial_\beta f^k} - \partial_{\bar\beta} f^l \overline{\partial_\alpha f^k}\right)}\omega^n$$

where we have used $R_{i\bar j k \bar l} = R_{i\bar l k \bar j}$, (a consequence of the first Bianchi identity) to exchange the indices j and l.

This motivates the following

<u>Def.</u>: *The curvature tensor* $R_{i\bar j k \bar l}$ *of* M *is called strongly (semi)negative if*

$$R_{i\bar j k \bar l}\left(A^i \overline{B^j} - C^i \overline{D^j}\right)\overline{\left(A^l \overline{B^k} - C^l \overline{D^k}\right)} > 0 \qquad (\geq 0)$$

for any $A^i, B^i, C^i, D^i \in \mathbb{C}$, *provided*

$$A^i \overline{B^j} - C^i \overline{D^j} \neq 0$$

for at least one pair (i,j).

This condition can be interpreted as follows (cf. [S4]):

Let R be the curvature tensor of M and put

$$H_R\left(X \otimes \bar Y\right) = R(X, \bar X, Y, \bar Y) = R(X, \bar Y, Y, \bar X)$$

for $X, Y \in T_a M$. H_R then induces a Hermitian form on $T^{1,0} M \otimes T^{0,1} M$. Then M has negative holomorphic bisectional curvature if

$$H_R \left(X \otimes \bar{Y} \right) > 0$$

for all $X, Y \neq 0 \in T_a M$,

negative sectional curvature if

$$H_R \left(X \otimes \bar{Y} - Y \otimes \bar{X} \right) > 0$$

for all such X, Y (the sectional curvature of the plane spanned by X and Y is given by

$$-\frac{1}{|X \wedge Y|^2} H_R \left(X \otimes \bar{Y} - Y \otimes \bar{X} \right),$$

cf. (5.2.2)),

and has strongly negative curvature if

$$H_R \left(X \otimes \bar{Y} + Z \otimes \bar{W} \right) > 0$$

for all nontrivial entries $(X, Y, Z, W \in T_a M)$.

Therefore, strongly negative curvature is a stronger condition than negative sectional or bisectional curvature.

We note

Lemma 5.2.3: If $f : N \to M$ is smooth and M has strongly seminegative curvature, then

(5.2.6) $$R_{i\bar{j}k\bar{l}} \, \bar{\partial} f^i \wedge \partial f^{\bar{j}} \wedge \partial f^k \wedge \overline{\partial f^l} \wedge \omega^{n-2} = b \omega^n$$

with a nonnegative function b on N.

As a consequence of the preceding three lemmas, we obtain

<u>Prop. 5.2.1:</u> Let N, M be Kähler manifolds, N compact, $f : N \to M$ harmonic.
If the curvature of M is strongly seminegative, then

(5.2.7)
$$R_{i\bar{j}k\bar{l}}\ \bar{\partial}f^i \wedge \partial f^{\bar{j}} \wedge \partial f^k \wedge \overline{\partial f^l} \equiv 0$$

and

(5.2.8)
$$D'\bar{\partial}f^i \equiv 0 \qquad\qquad \text{for all } i.$$

<u>pf.:</u> Since

$$\partial\bar{\partial}\left(g_{i\bar{j}}\bar{\partial}f^i \wedge \partial f^{\bar{j}} \wedge \omega^{n-2}\right) = \partial\bar{\partial}\left(g_{i\bar{j}}\bar{\partial}f^i \wedge \partial f^{\bar{j}}\right) \wedge \omega^{n-2}$$

is exact and N is compact, from Lemma 5.2.1,

$$\int_N \left(R_{i\bar{j}k\bar{l}}\ \bar{\partial}f^i \wedge \partial f^{\bar{j}} \wedge \partial f^k \wedge \overline{\partial f^l} \wedge \omega^{n-2} - g_{i\bar{j}}D'\bar{\partial}f^i \wedge D''\partial f^{\bar{j}} \wedge \omega^{n-2}\right) = 0,$$

and from Lemmas 5.2.2 and 5.2.3, we conclude that the integrand vanishes
pointwise. (5.2.7) then is clear, and (5.2.8) follows from the proof of Lemma
5.2.2.

$$\text{qed.}$$

<u>Note:</u> (5.2.7) implies the vanishing of the expression of the right hand side of (5.2.5).
Although (5.2.5) holds only if $\gamma_{\alpha\bar{\beta}} = \delta_{\alpha\beta}$, this expression then vanishes in any
holomorphic coordinate system, as the indices of the arguments are skew symmetric
in this expression so that nonorthonormal contributions always cancel.

The condition

$$D'\bar{\partial}f^i \equiv 0 \qquad\qquad \text{(for all } i)$$

means in local coordinates that

$$\partial_\alpha \partial_{\bar{\beta}} f^i + \Gamma^i_{jk}\ \partial_\alpha f^j\ \partial_{\bar{\beta}} f^k = 0 \qquad\qquad \text{(all } i)$$

for all α, β. In particular, the restriction of f to any local holomorphic curve or any local complex submanifold is again harmonic. We call such an f pluriharmonic. In particular, this condition is independent of the Kähler metric of N, but depends only on the holomorphic structure.

It is also instructive to derive a similar consequence for harmonic maps between noncompact Kähler manifolds (cf. [JY3] and [JY4]):

Prop. 5.2.2: Let $f : N \to M$ be a harmonic map of finite energy between Kähler manifolds N and M.

Suppose on N we have a family φ_ϵ of cut-off functions with the following properties:

(i) $0 \leq \varphi_\epsilon(z) \leq 1$ for all $z \in N$

(ii) φ_ϵ has compact support in N

(iii) $\varphi_\epsilon \to 1$ as $\epsilon \to 0$

(iv) $|\Delta\varphi_\epsilon(z)| \leq c$ (independent of z and ϵ)

Assume again that the curvature of M is strongly seminegative.

Then as in the preceding Proposition,

$$R_{i\bar{j}k\bar{l}}\, \bar{\partial}f^i \wedge \partial f^{\bar{j}} \wedge \partial f^k \wedge \overline{\partial f^l} \equiv 0$$

and

$$D'\bar{\partial}f^i \equiv 0$$

Note: The existence of the cut-off functions φ_ϵ is no serious restriction for applications as usually the metric of N can be modified to guarantee this assumption, cf. [JY3] and [JY4] for details.

pf.: First

$$\left| \int_N \varphi_\epsilon \partial\bar{\partial} \left(g_{i\bar{j}} \bar{\partial}f^i \wedge \partial f^{\bar{j}} \right) \wedge \omega^{n-2} \right| = \left| \int_N g_{i\bar{j}} \bar{\partial}f^i \wedge \partial f^{\bar{j}} \wedge \omega^{n-2} \wedge \partial\bar{\partial}\varphi_\epsilon \right|$$

$$\leq \int_N |df|^2 |\Delta\varphi_\epsilon|\, dvol\,(N)$$

$$\leq c\, E(f) \qquad\qquad\qquad\qquad \text{by (iv)}$$

Since by assumption $E(f) < \infty$, we can let $\varepsilon \to 0$ to obtain

$$\left| \int_N \partial \bar{\partial} \left(g_{i\bar{j}} \partial f^i \wedge \partial f^{\bar{j}} \right) \wedge \omega^{n-2} \right| < \infty$$

From Lemmas 5.2.1, 5.2.2 and 5.2.3, we conclude

$$\left| \int g_{i\bar{j}} D' \bar{\partial} f^i \wedge D'' \partial f^{\bar{j}} \wedge \omega^{n-2} \right| < \infty$$

On the other hand, from the proof of Lemma 5.2.2, since f is harmonic

$$g_{i\bar{j}} D' \bar{\partial} f^i \wedge D'' \partial f^{\bar{j}} \wedge \omega^{n-2} = -\frac{4}{n^2 - n} \left| D' \bar{\partial} f \right|^2 \omega^n ,$$

hence

(5.2.8)
$$\int \left| D' \bar{\partial} f \right|^2 \, \mathrm{dvol}\,(N) < \infty$$

Therefore, we can find an exhaustion of N by an increasing sequence (N_l) of smooth subsets for which

$$\left| \int_{N_l} \partial \bar{\partial} \left(g_{i\bar{j}} \bar{\partial} f^i \wedge \partial f^{\bar{j}} \right) \wedge \omega^{n-2} \right| = \left| \int_{\partial N_l} \bar{\partial} \left(g_{i\bar{j}} \bar{\partial} f^i \wedge \partial f^{\bar{j}} \right) \wedge \omega^{n-2} \right|$$

$$\leq \left(\int_{\partial N_l} |df|^2 \right)^{\frac{1}{2}} \left(\int_{\partial N_l} \left| D' \bar{\partial} f \right|^2 \right)^{\frac{1}{2}}$$

(for this step, note that $g_{i\bar{j},k} \bar{\partial} f^k \wedge \bar{\partial} f^i \wedge f^{\bar{j}} = 0$ as $g_{i\bar{j},k} = g_{k\bar{j},i}$ and $\bar{\partial} f^k \wedge \bar{\partial} f^i = -\bar{\partial} f^i \wedge \bar{\partial} f^k$, and furthermore that $|df|^2$ involves one and $|D' \bar{\partial} f|^2$ two contractions with $\gamma^{\alpha \bar{\beta}}$, the inverse domain metric)

$$\to 0 \qquad\qquad \text{as } l \to \infty \text{ (by choice of } N_l)$$

Hence

$$\int_N \partial \bar{\partial} \left(g_{i\bar{j}} \bar{\partial} f^i \wedge \partial f^{\bar{j}} \right) \wedge \omega^{n-2} = 0,$$

and we can argue as in the preceding proof.

Of course, since we are now working with the form $g_{i\bar{j}} \bar{\partial} f^i \wedge \partial f^{\bar{j}}$ instead of the function $|df|^2$, a Bochner type identity does not lead to pointwise estimates anymore as in chapter 3. From the preceding proof, however, we see that it can still be useful to obtain integral estimates such as (5.2.8).

__Prop. 5.2.3:__ *Suppose* $f : N \to M$ *is harmonic,* $\mathrm{rank}_R \, df \geq 3$ *at some* $p \in N$,

(5.2.9)
$$\partial_\alpha f^i \overline{\partial_\beta f^j} - \partial_{\bar{\beta}} f^i \overline{\partial_\alpha f^j} \equiv 0 \qquad \qquad in \ N.$$

Then f *is either holomorphic or antiholomorphic in* N.

pf.: By the lower semicontinuity of the rank,

$$\mathrm{rank}_R \, df \geq 3 \qquad \qquad \text{in some open } U \in N.$$

We first show that for all $p \in U$, either

$$\partial f(p) = 0$$

or
$$\bar{\partial} f(p) = 0$$

If

(5.2.10)
$$\partial f^j(p) \neq 0,$$

then for some α, $\partial_\alpha f^j(p) \neq 0$, hence from (5.2.9)

$$\partial_{\bar{\beta}} f^i = \frac{\partial_\alpha f^i}{\overline{\partial_\alpha f^j}} \, \overline{\partial_\beta f^j} \qquad \qquad \text{(note that no summation is involved),}$$

i.e.

$$\bar{\partial} f^i \text{ is linearly dependent of } \overline{\partial f^j} \text{ over } \mathbb{C}$$

for each i.

If

(5.2.11)
$$\bar{\partial} f^i(p) \neq 0 \qquad \qquad \text{for some } i,$$

then consequently all $\overline{\partial f^k}(p)$ are linearly dependent of $\bar{\partial} f^i(p)$ over $\mathbb{C}$. Since also all $\bar{\partial} f^i$ are linearly dependent of $\overline{\partial f^j}(p)$ over $\mathbb{C}$, the real rank of $df(p)$ is at most 2, contradicting $p \in U$. Hence (5.2.10) and (5.2.11) cannot hold together.

Thus, for each $p \in U$, either $\partial f(p) = 0$ or $\bar{\partial} f(p) = 0$.

On the other hand, since $\text{rank}_R \, df > 0$ in U, $\partial f(p)$ and $\bar{\partial} f(p)$ cannot vanish simultaneously at $p \in U$, and hence U has to coincide with one of its closed subsets

$$\{p \in U : \partial f(p) = 0\}$$

or

$$\{p \in U : \bar{\partial} f(p) = 0\},$$

i.e. either

$$\partial f \equiv 0 \qquad \text{in } U$$

or

$$\bar{\partial} f \equiv 0 \qquad \text{in } U.$$

The claim then follows from

<u>Prop. 5.2.4:</u> *Let again* $f : N \to M$ *be harmonic, and let* N *be connected,* U *a nonempty open subset of* N. *If* f *is (anti)holomorphic on* U, *then* f *is (anti)holomorphic in* N.

pf.: Let e.g.

$$\bar{\partial} f \equiv 0 \qquad \text{in } U$$

Let U' be the largest connected open subset of N containing U with

$$\bar{\partial} f \equiv 0 \qquad \text{in } U'$$

IF $U' \neq N$, then there exists $p \in \partial U'$. Let W be a connected open neighbourhood of p, so small that W and $f(W)$ can both be covered by single local holomorphic coordinate charts defined on a neighbourhood of the choice of W and $f(W)$, resp.

We differentiate the harmonic map equation in local coordinates

$$\gamma^{\alpha\bar{\beta}} \partial_{\bar{\beta}} \partial_\alpha f^i + \gamma^{\alpha\bar{\beta}} \, \Gamma^i_{jk} \, \partial_\alpha f^j \, \partial_{\bar{\beta}} f^k = 0$$

w.r.t. ∂_η and obtain

$$\left| \Delta \left(\partial_\eta f^i \right) \right| \leq c \left(\sum_{\alpha, j} \left(\left| \text{grad} \, \partial_\alpha f^i \right| + \left| \partial_\alpha f^j \right| \right) \right)$$

(putting for example all factors $\partial_\alpha f^j$ into the constant c)

($\Delta = \Delta_N = \gamma^{\alpha\bar\beta} \partial_{\bar\beta} \partial_\alpha$ is the Laplace–Beltrami operator of N). Let $u^{i,\eta}$, $v^{i,\eta}$ be real and imaginary part, resp., of $\partial_\eta f^i$. Since Δ is real,

$$\left| \Delta \left(\partial_\eta f^i \right) \right| = \left| \Delta u^{i,\eta} \right| + \left| \Delta v^{i,\eta} \right|,$$

hence

$$\left| \Delta u^{i,\eta} \right|^2 \leq c \sum_{\alpha,j} \left(\left| \text{grad } u^{j,\alpha} \right|^2 + \left| \text{grad } v^{j,\alpha} \right|^2 + \left| u^{j,\alpha} \right|^2 + \left| v^{j,\alpha} \right|^2 \right)$$

$$\left| \Delta v^{i,\eta} \right|^2 \leq c \sum_{\alpha,j} \left(\left| \text{grad } u^{j,\alpha} \right|^2 + \left| \text{grad } v^{j,\alpha} \right|^2 + \left| u^{j,\alpha} \right|^2 + \left| v^{j,\alpha} \right|^2 \right)$$

Aronszajn's unique continuation theorem implies

$$u^{i,\eta} \equiv 0 \equiv v^{i,\eta}$$

on W, since this is true on $W \cap U'$. Hence by definition of U', $p \in U'$, and consequently U' cannot have a boundary point, i.e. $U' = N$.

qed.

Thm. 5.2.1: $g : N \to M$ _a continuous map of compact Kähler manifolds, M strongly negatively curved,_

(5.2.12) $$g_* : H_l(N, \mathbb{R}) \to H_l(M, \mathbb{R})$$

nontrivial for some $l \geq 3$.

Then g is homotopic to a holomorphic or antiholomorphic map.

pf.: By Thm. 3.3.1, g is homotopic to a harmonic map f. By (5.2.12), $\text{rank}_\mathbb{R} \, df \geq 3$ somewhere. Hence f is (anti)holomorphic by Prop. 5.2.3.

Cor. 5.2.1: _M compact Kähler manifold of strongly negative curvature. Then any element of $H_{2k}(M, \mathbb{Z})$ ($k \geq 2$) that can be represented by the continuous image of a compact Kähler manifold can also be represented by a complex-analytic subvariety of M._

<u>Thm. 5.2.2</u>: Let $g : N \to M$ be a continuous map between compact Kähler manifolds. Assume M has strongly negative sectional curvature.

Suppose $\dim_{\mathbb{C}} M = \dim_{\mathbb{C}} N = n \geq 2$, $\deg g = 1$, and

$$g_* : H_{2n-2}(N, \mathbb{R}) \to H_{2n-2}(M, \mathbb{R})$$

is injective. Then g is homotopic to a biholomorphic map from N to M.

<u>pf.</u>: By Thm. 5.2.1, g is homotopic to a holomorphic f (noting $\deg f = \deg g = 1$).

Let $V := \{p \in N : f \text{ not a local homeomorphism at } p\}$.

Since $\deg f = \deg g = 1$, $V \neq N$.

If $V \neq \emptyset$, then it is a subvariety of N of complex codimension 1, since locally defined by

$$\det \left(\frac{\partial f^i}{\partial z^\alpha} \right) = 0$$

By the injectivity of f_* on H_{2n-2}, $f(V)$ is also (a subvariety of M) of complex codimension 1.

Hence there exists $p \in V$ which is isolated in $f^{-1}(f(p))$ and for which $f^{-1}|_{f(V)}$ exists in a neighbourhood of $f(p)$ as a well-defined map.

Since $N - f^{-1}(f(V))$ is mapped bijectively onto $M - f(V)$, because of $\deg f = 1$, we can choose local holomorphic coordinates z^α at p and remove the isolated singularity of $z^\alpha \cdot f^{-1}$ at $f(p)$ $(\alpha = 1, \dots, n)$.

Hence f is locally diffeomorphic at p, contradicting $p \in V$. Hence $V = \emptyset$.

qed.

<u>Cor.5.2.2</u>: M a compact Kähler manifold, $\dim_{\mathbb{C}} M \geq 2$, strongly negatively curved. Then any compact Kähler manifold N of the same homotopy type as M is (anti)biholomorphic to M.

<u>pf.</u>: After possibly changing the orientation of N, we find a continuous homotopy equivalence $g : N \to M$ of degree 1 and apply the previous result.

qed.

For applications, we note

Prop. 5.2.4: *The curvature tensor of the Hermitian symmetric metric of the unit ball in $\mathbb{C}^n$ $(n \geq 2)$ is strongly negative. Hence Cor. 5.2.2 applies to compact quotients (without singularities) of the unit ball.*

pf.: The Bergmann kernel function is

$$\log \frac{1}{1 - |z|^2},$$

hence the Kähler metric given by

$$g_{i\bar{j}} = \partial_i \partial_{\bar{j}} \log \frac{1}{1 - |z|^2}$$

Because of the symmetric structure, we have to check the curvature condition only at 0. Since

$$\log \frac{1}{1 - |z|^2} = |z|^2 + \frac{|z|^4}{2} + O(|z|^6),$$

$$R_{i\bar{j}k\bar{l}}(0) = \partial_i \partial_{\bar{j}} g_{k\bar{l}}(0) = \frac{1}{2} \partial_i \partial_{\bar{j}} \partial_k \partial_{\bar{l}} |z|^4,$$

and the condition is easily checked.

Remark: A different example of a strongly negatively curved Kähler manifold was constructed by Mostow-Siu [MS].

A more careful analysis of the curvature tensor shows that for irreducible Hermitian symmetric domains of complex dimension at least 2, one can still show that for harmonic maps of sufficiently high rank, the vanishing of

$$R_{i\bar{j}k\bar{l}} \, \bar{\partial} f^i \wedge \partial f^j \wedge \partial f^k \wedge \overline{\partial f^l}$$

implies that f is (anti)holomorphic. In particular, one gets a strong rigidity result as in Cor. 5.2.2 for compact quotients (without singularities) of such domains, cf. [S1], [S2], [Zh], [S3] for details.

We now look at harmonic maps $f : N \to \Sigma$ from a compact Kähler manifold N into a compact regular curve, i.e. a compact Riemann surface without singularities, of genus at least 2, equipped with the standard hyperbolic metric of curvature -1. The curvature of Σ is then strongly negative, but one cannot expect f to be (anti)holomorphic; let e.g. $N = \Sigma_1$ where Σ_1 is a Riemannian surface diffeomorphic to, but not conformally equivalent to Σ. Then in every homotopy class we can find a harmonic map f by Thm. 3.3.1, but since Σ_1 and Σ have different conformal structures, f cannot be holomorphic. The same of course is true for higherdimensional domains, e.g. $N = \Sigma_1 \times \Sigma_2$, Σ_1 as before, Σ_2 an arbitrary Riemann surface. If $f : \Sigma_1 \to \Sigma$ is harmonic, but not holomorphic, $\pi : \Sigma_1 \times \Sigma_2 \to \Sigma_1$ the projection, then $f \circ \pi : N \to \Sigma$ is harmonic, but not holomorphic. Likewise by taking products, one can also construct higher dimensional images with this property. Nevertheless, one has

<u>Thm. 5.2.3</u> ([JY1]): Let $f : N \to \Sigma$ be harmonic, N a compact Kähler manifold, Σ a compact Riemann surface (without singularities) with hyperbolic metric. If $z_0 \in N$ and $df(z_0) \neq 0$, then there exists a neighbourhood U of z_0 so that the local level sets $\{f \equiv c\} \cap U$ are unions of analytic hypersurfaces.

<u>pf.</u>: Let the metric of Σ in local coordinates be given by

$$\rho^2(f)\ df\ d\bar{f}$$

The corresponding Christoffel symbol then is $\Gamma = \frac{2\rho_f}{\rho}$. Let, as before, $(\gamma_{\alpha\bar{\beta}})$ be the metric tensor of N in local holomorphic coordinates. The harmonic map equation then is

$$\gamma^{\alpha\bar{\beta}}\left(\partial_{\bar{\beta}}\partial_\alpha f + \Gamma\ \partial_\alpha f\ \partial_{\bar{\beta}} f\right) = 0$$

By Prop. 5.2.1, f is pluriharmonic, i.e.

(5.2.13) $$\partial_{\bar{\beta}}\partial_\alpha f + \Gamma\ \partial_\alpha f\ \partial_{\bar{\beta}} f = 0 \qquad \text{for all } \alpha, \beta$$

Furthermore, since the curvature of Σ is negative

(5.2.14) $$\partial_\alpha f\ \overline{\partial_\beta f} - \partial_{\bar{\beta}} f\ \overline{\partial_\alpha f} = 0 \qquad \text{for all } \alpha, \beta.$$

Hence, if e.g. $\partial_\alpha f(z_0) \neq 0$, then in a neighbourhood U of z_0

(5.2.15)
$$\bar{\partial} f(z) = \lambda(z) \, \overline{\partial f}(z)$$

with $\lambda(z) = \dfrac{\partial_\alpha f(z)}{\overline{\partial_\alpha f}(z)}$, i.e. $\bar{\partial} f(z)$ and $\overline{\partial f}(z)$ are linearly dependent over $\mathbb{C}$.
If $\partial_\alpha f(z_0) \neq 0$, we can put, again in a neighbourhood U of z_0

(5.2.16)
$$\partial_\beta f(z) = k_\beta(z) \partial_\alpha f(z) \qquad \text{for each } \beta$$

From (5.2.13)

$$0 = \partial_\eta \partial_\beta f + \Gamma \, \partial_\beta f \, \partial_\eta f$$

$$= (\partial_\eta k_\beta) \partial_\alpha f + k_\beta (\partial_\eta \partial_\alpha f + \Gamma \, \partial_\alpha f \, \partial_\eta f)$$

Hence, using again (5.2.13) and $\partial_\alpha f \neq 0$ in U,

$$\partial_\eta k_\beta = 0$$

Thus, the k_β are holomorphic.

We now want to find holomorphic coordinates $\xi^1, \ldots, \xi^n$ in U with

$$\frac{\partial f}{\partial \xi^\beta} = 0 \qquad \text{for all } \beta \neq \alpha.$$

We have to solve

(5.2.17)
$$\frac{\partial f}{\partial z^\eta} \frac{\partial z^\eta}{\partial \xi^\beta} = 0 \qquad \text{for } \beta \neq \alpha$$

From (5.2.16) we get, since $\frac{\partial f}{\partial z^\alpha} \neq 0$ in U,

$$\frac{\partial z^\alpha}{\partial \xi^\beta} + \sum_{\eta \neq \alpha} k_\eta \frac{\partial z^\eta}{\partial \xi^\beta} = 0. \qquad (\beta \neq \alpha).$$

This system of equations now is solvable by the complex version of Frobenius' Theorem since the k_η are holomorphic[1] and the distribution given by the vectors

$$e^\beta = (e_1^\beta, \ldots, e_n^\beta) \qquad (\beta \neq \alpha)$$

This avoids using the Newlander-Nirenberg Theorem [NN].

with

$$e_\alpha^\beta = k_\beta \ , \qquad e_\beta^\beta = -1 \ , \qquad e_\gamma^\beta = 0 \qquad \text{for } \gamma \neq \alpha, \beta$$

is integrable, i.e. for $\beta, \delta \neq \alpha$

$$[e^\beta, e^\delta] = k_\beta \frac{\partial k_\delta}{\partial z^\alpha} - \frac{\partial k_\delta}{\partial z^\beta} - k_\delta \frac{\partial k_\beta}{\partial z^\alpha} + \frac{\partial k_\beta}{\partial z^\delta}$$

$$= 0 \qquad \qquad \text{by } (5.2.16).$$

From (5.2.15) we see that for $\beta \neq \alpha$

$$\frac{\partial f}{\partial \xi^\beta} = \frac{\partial f}{\partial z^{\bar{\gamma}}} \frac{\partial z^{\bar{\gamma}}}{\partial \xi^\beta} = \lambda \frac{\overline{\partial f}}{\partial z^\gamma} \cdot \frac{\overline{\partial z^\gamma}}{\partial \xi^\beta} = 0 \qquad \qquad \text{by } (5.2.17).$$

Hence f is independent of ξ^β for $\beta \neq \alpha$, i.e. the level sets of f are unions of analytic hypersurfaces $\xi^\alpha = \text{const.}$

qed.

At points where $df(z) = 0$, a more careful analysis is needed. With the help of the holomorphic foliation of the preceding theorem, one can prove ([JY2] for $n = 2$, [M1], [M2] for the general case, cf. also [S4] for a simplified proof), again by showing that a harmonic map exhibits the required properties:

Thm. 5.2.4: Let M, a compact Kähler manifold, be of the form Δ^m/Γ where Δ is the hyperbolic unit disk and $\Gamma \subset \text{Aut}(\Delta^m)$ is discrete and fixpoint free. If N is another compact Kähler manifold which is homotopically equivalent to M, then M is also of the form Δ^m/Γ', where Γ' (as an abstract group) is isomorphic to Γ. If Γ is irreducible, i.e. no finite cover of Δ^m/Γ splits as a product, then $\Gamma' = \Gamma$ as a subgroup of $\text{Aut}(\Delta^m)$ (after possibly changing the orientation of N), and M and N are biholomorphically equivalent.

Another consequence of the preceding analysis is (cf. also [S4])

<u>Thm. 5.2.5:</u> N a compact Kähler manifold, Σ a compact hyperbolic Riemann surface,
$g : N \to \Sigma$ continuous, $g_* : H_2(N, \mathbb{R}) \to H_2(\Sigma, \mathbb{R})$ nontrivial.

Then there exist a compact hyperbolic Riemann surface Σ', a holomorphic
map $h : N \to \Sigma'$, a harmonic map $\varphi : \Sigma' \to \Sigma$ so that the harmonic map
$f = \varphi \circ h$ is homotopic to g.

In particular, the lifting $\tilde{h} : \tilde{N} \to \Delta$ to universal covers is a nonconstant
bounded holomorphic function on $\tilde{N}$.

Some remarks on notation and terminology

In Riemannian geometry, there exist some competing sign conventions. The Laplacian can be defined on forms as

$$\Delta^+ = dd^* + d^*d$$

where d^* is the L^2-adjoint of d. This is a positive operator, i.e.

$$(\Delta^+\omega, \omega) = (d\omega, d\omega) + (d^*\omega, d^*\omega) \geq 0,$$

and all eigenvalues are nonnegative. This is the geometric Laplacian. One also has the analytic Laplacian, defined on functions in $\mathbb{R}^m$ as

$$\Delta^- f = \sum_{i=1}^m \frac{\partial^2 f}{\partial x_i^2},$$

and its generalization to Riemannian manifolds, the Laplace-Beltrami operator Δ^-. This is a negative operator. Δ^- is the Laplace operator commonly used by analysts; e.g. a function f is subharmonic if

$$\Delta^- f \geq 0.$$

On functions on a manifold, one has

$$\Delta^- = -\Delta^+$$

We shall use both Laplace operators; in a given context usually the one that seems most natural in that situation. We always distinguish them by the superscript $^+$ or $^-$ so that no confusion is possible.

There are are also two different sign conventions for the curvature tensor. We shall use the one of Kobayashi-Nomizu ([KN]). If X and Y are orthonormal tangent vectors at $x \in M$ (M a Riemannian manifold), then the sectional curvature of the tangent plane spanned by X and Y is given by

$$K(X,Y) = R(X,Y,Y,X);$$

this convention is different from the one e.g. used in [Kl] and [J2].

If, however, $e_1, \ldots, e_m$ is an orthonormal basis of $T_x M$, $X = x^i e_i$, $Y = y^i e_i$, then in tensor notation

$$K(X, Y) = R_{ijkl}\, x^i y^j x^k y^l,$$

and here both sign conventions agree.

Note that in the preceding discussion, we already used the Einstein summation convention, i.e. to sum an index occuring twice in a product over its range.

For example

$$x^i e_i = \sum_{i=1}^{m} x^i e_i.$$

We shall denote a pointwise scalar product by $< \cdot, \cdot >$ and a pointwise norm by $| \cdot |$. L^2-products and norms will be denoted by $(\cdot, \cdot)$ and $\| \cdot \|$; for example, for $\omega \in \Omega^p(M)$, i.e. a p-form on M,

$$\|\omega\| = \left(\int_M \omega \wedge *\omega \right)^{\frac{1}{2}}$$

where $*$ is defined in 1.2.

A distance function is denoted by $d(\cdot, \cdot)$, and for $x \in M$, $r > 0$

$$B(x, r) := \{ y \in M : d(x, y) \le r \}$$

is a closed ball of radius r.

Occasionally, we shall mention Sobolev spaces $H^{k,p}$. $H^{k,p}$ is the completion of C_0^∞ (compactly supported C^∞-functions) w.r.t. the norm

$$\|f\|_{H^{k,p}} := \sum_{j=1}^{k} \left(\int |\nabla^j f|^p \right)^{\frac{1}{p}},$$

where $\nabla^j f$ denotes the tensor of the jth (covariant) derivatives of f.

We shall however obtain all our estimates in the Hölder spaces $C^{k,\alpha}$ instead. Here $k \in \{0, 1, \ldots\}$, and $\alpha \in (0, 1)$, and $C^{k,\alpha}$ is the space of k-times differentiable functions, the kth derivatives of which satisfy a Hölder condition with exponent α. If we speak of $C^{k,\alpha}$ estimates then we usually mean that they are valid for the given k and for all $\alpha \in (0,1)$; the constants in the estimates may depend on α.

We also note that, by the Arzela-Ascoli theorem, the embedding $C^{k,\alpha} \to C^{k,\beta}$ with $\beta < \alpha$ is compact.

Finally, in an estimate, c will denote a constant depending on various geometric data but not on the estimated function, e.g.

$$\|f\|_{C^{p,\alpha}} \leq c\|\Delta f\|_{L^\infty}$$

and c is independent of f. In different estimates, we shall denote the constant by the same letter c as customary in analysis, although the value of c may change.

Bibliography

[Al1]   Al'ber, S.I., On n-dimensional problems in the calculus of variations in the large, Sov. Math. Dokl. **5** (1964), 700–704

[Al2]   Al'ber, S.I., Spaces of mappings into a manifold with negative curvature, Sov. Math. Dokl. **9** (1967), 6–9

[AHS]   Atiyah, M., N. Hitchin, and I. Singer, Self duality in four dimensional Riemannian geometry, Proc. Roy. Soc. London Ser. A **362** (1978), 425–461

[A]   Aubin, T., Nonlinear analysis on manifolds. Monge-Ampère equations, Springer, New York etc., 1982

[BK]   Buser, P., and H. Karcher, Gromov's Almost Flat Manifolds, Astérisque **81** (1981)

[CE]   Cheeger, J., and D. Ebin, Comparison theorems in Riemannian Geometry, Amsterdam, Oxford, 1975

[D1]   Donaldson, S., A new proof of a theorem of Narasimhan-Seshadri, J. Diff. Geom. **18** (1983), 269–277

[D2]   Donaldson, S., An application of gauge theory to four dimensional topology, J. Diff. Geom. **18** (1983), 279–315

[D3]   Donaldson, S., Anti self-dual Yang-Mills connections over complex algebraic surfaces and stable vector bundles, Proc. London Math. Soc. **50** (1985), 1–26

[D4]   Donaldson, S., Infinite determinants, stable bundles, and curvature, Duke Math. **54** (1987), 231–247

[EL]   Eells, J., and L. Lemaire, Selected Topics in Harmonic Maps, CBMS Regional Conf., 1981

[ES]   Eells, J., and J.H. Sampson, Harmonic Mappings of Riemannian Manifolds, Am. J. Math. **86** (1964), 109–160

[EW]   Eells, J., and J.C. Wood, Restrictions on Harmonic Maps of Surfaces, Top. **15** (1976), 263–266

[FU]   Freed, D., and K. Uhlenbeck, Instantons and four-manifolds, Springer, New York, etc., 1984

[GT]   Gilbarg, D., and N.S. Trudinger, Elliptic Partial Differential Equations of Second Order, Springer, Grundlehren 224, Berlin, Heidelberg, New York, 1977

[GH]   Griffiths, P., and J. Harris, Principles of algebraic geometry, Wiley, New York, 1978

[GW]   Gromoll, D., and J. Wolf, Some relations between the metric structure and the algebraic structure of the fundamental group in manifolds of nonpositive curvature, Bull. AMS **77** (1971), 545–552

[G]   Grove, K., Metric differential geometry, Lecture notes from the Nordic Summer School, Lyngby, 1985, in: V.L. Hansen (ed.), Differential Geometry, Springer Lecture Notes Math., Berlin etc. (1987), 171–227

[Hm]   Hamilton, R., Harmonic Maps of Manifolds with Boundary, L.N.M. 471, Springer, Berlin, Heidelberg, New York, 1975

[Ht]   Hartman, P., On Homotopic Harmonic Maps, Can. J. Math. **19** (1967), 673–687

[J1]   Jost, J., Harmonic maps between surfaces, LNM 1062, Springer, Berlin, etc., 1984

[J2]   Jost, J., Harmonic mappings between Riemannian manifolds, ANU-Press, Canberra, 1984

[J3]   Jost, J., Two dimensional geometric variational problems, Wiley – Interscience, Chichester, 1991

[JY1]   Jost, J., and S.T. Yau, Harmonic Mappings and Kähler Manifolds, Math. Ann. **262** (1983), 145–166

[JY2]   Jost, J., and S.T. Yau, A strong rigidity theorem for a certain class of compact analytic surfaces, Math. Ann. **271** (1985), 143–152

152

[JY3]   Jost, J., and S.T. Yau, The strong rigidity of locally symmetric complex manifolds of rank one and finite volume, Math. Ann., 275 (1986), 291–304

[JY4]   Jost, J., and S.T. Yau, On the rigidity of certain discrete groups and algebraic varieties, Math. Ann. **278** (1987), 481–496

[Kl]   Klingenberg, W., Riemannian geometry, de Gruyter, Berlin, New York, 1982

[Kb1]   Kobayashi, S., Curvature and stability of vector bundles, Proc. Japan. Acad. Math. Sci. **58** (1982), 158–162

[Kb2]   Kobayashi, S., Differential geometry of holomorphic vector bundles, Iwanami Shoten and Princeton University Press, 1987

[KN]   Kobayashi, S., and K. Nomizu, Foundations of differential geometry, 2 vols., Wiley, New York, 1963 and 1969

[LSU]   Ladyženskaja, O.A., V.A. Solonnikov, and N.N. Ural'ceva, Linear and quasilinear equations of parabolic type, AMS, Providence, R.I., 1968

[LY]   Lawson, B., and S.T. Yau, Compact manifolds of nonpositive curvature, J. Diff. Geom. **7** (1972), 211–228

[Lb]   Lübke, M., Stability of Einstein-Hermitian vector bundles, Man. math. **42** (1983), 245–257

[MR]   Milgram, A., and P. Rosenbloom, Harmonic forms and heat conduction, I: Closed Riemannian manifolds, Proc. Nat. Acad. Sci. **37** (1951), 180–184

[M1]   Mok, N., The holomorphic or antiholomorphic character of harmonic maps into irreducible compact quotients of polydisks, Math. Ann. 272 (1985), 197–216

[M2]   Mok, N., Strong rigidity of irreducible quotients of poydiscs of finite volume, Math. Ann. **282** (1988), 555–578

[M]   Morrey, C.B., Multiple Integrals in the Calculus of Variations, Springer, Berlin, Heidelberg, New York, 1966

[MS]   Mostow, G., and Y.T. Siu, A compact Kähler surface of negative curvature not covered by the ball, Ann. Math. **112** (1980), 321–360

[NN]   Newlander, A., and L. Nirenberg, Complex analytic coordinates in almost complex manifolds, Ann. Math. **65** (1957), 391–404

[Sa]   Sampson, J., Applications of harmonic maps to Kähler geometry, Contemp. Math. **49** (1986), 125–134

[S]   Schoen, R., Analytic aspects of the harmonic map problem, in S.S. Chern (ed.), Seminar on nonlinear partial differential equations, Springer, New York, etc., 1984

[SU]   Schoen, R., and K. Uhlenbeck, A Regularity Theory for Harmonic Maps, J. Diff. Geom. **17** (1982), 307–335

[SY]   Schoen, R., and S.T. Yau, Compact Group Actions and the Topology of Manifolds with Non-Positive Curvature, Top. **18** (1979), 361–380

[S1]   Siu, Y.T., The Complex Analyticity of Harmonic Maps and the Strong Rigidity of Compact Kähler Manifolds, Ann. Math. 112 (1980), 73–111

[S2]   Siu, Y.T., Strong rigidity of compact quotients of exceptional bounded symmetric domains, Duke Math. J. **48** (1981), 857–871

[S3]   Siu, Y.T., Complex-analyticity of harmonic maps, vanishing and Lefschetz theorems, J. Diff. Geom. **17** (1982), 55–138

[S4]   Siu, Y.T., Strong rigidity for Kähler manifolds and the construction of bounded holomorphic functions, in: R. Howe (ed.), Discrete groups in geometry and analysis, 124–151, Birkhäuser, Boston, 1987

[T]   Tian, G., On Kähler-Einstein metrics on certain Kähler manifolds with $c_1(M) > 0$, Inv. math. **89** (1987), 225–246

[U1]   Uhlenbeck, K., Removable singularities in Yang-Mills fields, Comm. Math. Phys. **83** (1982), 11–29

[U2]   Uhlenbeck, K., Connections with L^p-bounds on curvature, Comm. Math. Phys. **83** (1982), 31–42

[UY]   Uhlenbeck, K., and S.T. Yau, On the existence of Hermitian Yang-Mills connections in stable vector bundles, CPAM 39, No. 5 (1986), 257–293

[W]   Wells, Differential analysis on complex manifolds, Springer, Berlin, etc., ^{2}1980

[Wu]   Wu, H.-H., The Bochner technique in differential geometry

[Y1]   Yau, S.T., Calabi's conjecture and some new results in algebraic geometry, Proc. NAS USA **74** (1977), 1798–1799

[Y2]   Yau, S.T., On the Ricci curvature of a compact Kähler manifold and the complex Monge-Ampère equation, I. CPAM **31** (1978), 339–411

[Z]   Zhong, J.-Q., The degree of strong nondegeneracy of the bisectional curvature of exceptional bounded symmetric domains, in: Proc. Int. Conf. Several Complex Variables, Kohn, Lu, Remmert, Siu (eds.), Birkhäuser, Boston, 1984, 127–139

Previously published in the series DMV Seminar:

Volume 1: **Manfred Knebusch/Winfried Scharlau, Algebraic Theory of Quadratic Forms.**
1980, 48 pages, softcover, ISBN 3-7643-1206-8.

Volume 2: **Klas Diederich/IngoLieb, Konvexitaet in der Komplexen Analysis.**
1980, 150 pages, softcover, ISBN 3-7643-1207-6.

Volume 3: **S. Kobayashi/H. Wu/C. Horst, Complex Differential Geometry.**
2nd edition 1987, 160 pages, softcover, ISBN 3-7643-1494-X.

Volume 4: **R. Lazarsfeld/A. van de Ven, Topics in the Geometry of Projective Space.**
1984, 52 pages, softcover, ISBN 3-7643-1660-8.

Volume 5: **Wolfgang Schmidt, Analytische Methoden für Diophantische Gleichungen.**
1984, 132 pages, softcover, ISBN 3-7643-1661-6.

Volume 6: **A. Delgado/D. Goldschmidt/B. Stellmacher, Groups and Graphs: New Results and Methods.**
1985, 244 pages, softcover, ISBN 3-7643-1736-1.

Volume 7: **R. Hardt/L. Simon, Seminar on Geometric Measure Theory.**
1986, 118 pages, softcover, ISBN 3-7643-1815-5.

Volume 8: **Yum-Tong Siu, Lectures on Hermitian-Einstein Metrics for Stable Bundles and Kaehler-Einstein Metrics.**
1987, 172 pages, softcover, ISBN 3-7643-1931-3.

Volume 9: **Peter Gaenssler/Winfried Stute, Seminar on Empirical Processes.** 1987, 114 pages, softcover, ISBN 3-7643-1921-6.

Volume 10: **Jürgen Jost, Nonlinear Methods in Riemannian and Kaehlerian Geometry.** 1988, 154 pages, softcover, ISBN 3-7643-1920-8.

Volume 11: **Tammo tom Dieck/Ian Hambleton, Surgery Theory and Geometry of Representations.**
1988, 122 pages, softcover, ISBN 3-7643-2204-7.